D1031819

THE PHILOSOPHY OF MATHEMATICS

The Invisible Art

THE PHILOSOPHY OF MATHEMATICS

The Invisible Art

W.S. Anglin

Studies in the History of Philosophy
Volume 43

The Edwin Mellen Press
Lewiston/Queenston/Lampeter

Library of Congress Cataloging-in-Publication Data

Anglin, W. S.
  The philosophy of mathematics : the invisible art / W.S. Anglin.
     p.   cm. -- (Studies in the History of philosophy ; v. 43)
  Includes bibliographical references and index.
  ISBN 0-7734-8706-9 (hard)
  1. Mathematics--Philosophy.   I. Title. II. Series: Studies in
the history of philosophy (Lewiston, N.Y.) ; v. 43.
  QA8.4.A54 1997
  510'.1--dc21   96-48604
                                                                  CIP

This is volume 43 in the continuing series
Studies in the History of Philosophy
Volume 43 ISBN 0-7734-8706-9
SHP Series ISBN 0-88946-300-X

A CIP catalog record for this book is available from the British Library.

The Edwin Mellen Press   The Edwin Mellen Press
        Box 450   Box 67
Lewiston, New York   Queenston, Ontario
 USA 14092-0450   CANADA L0S 1L0

The Edwin Mellen Press, Ltd.
Lampeter, Dyfed, Wales
UNITED KINGDOM SA48 7DY

Printed in the United States of America

To Diana Zhang

Table of Contents

Preface

Unlike ethics, the philosophy of mathematics does not address life or death issues. No one has been imprisoned for holding the wrong view about the Axiom of Choice. None the less, the philosophy of mathematics is related to basic topics in metaphysics, epistemology, and values, and one can explore those topics via a consideration of the ontology, knowledge or worth of mathematical objects. Everything in this book is linked to large, nonmathematical questions.

Chapters 1 and 2 give a brief history of what different philosophers have said about infinity, and Chapter 3 presents some arguments for the position that one should adopt a pro-infinity attitude. Chapter 4 introduces the ontological 'schools' (viz. realism, intuitionism, and formalism), and Chapter 5 examines the relationship between mathematics and various theories of truth. Chapters 6 and 7 explore mathematics and values. One of the questions considered is 'what makes an elegant proof elegant?' Chapter 8 focuses on presuppositions in the history of mathematics. Is it an epic in which Reason defeats Superstition — or is it a comedy of errors? Chapter 9 looks at mathematical knowledge, analysing it in terms of some traditional epistemological distinctions, and Chapter 10 investigates the role of mathematics in education. Most philosophers agree that a good education includes mathematics. Chapter 11 highlights some of the implications which orthodox theism has for the philosophy of mathematics, and, finally, Chapter 12 deals with the problem of referring to mathematical objects.

Some philosophy of mathematics is written on the assumption that the reader is a specialist in modal logic, or descriptive set theory, or category theory, or philosophy of language. This book is less technical. It is meant to give an overview of central issues, an overview that will be accessible to anyone involved in either mathematics or philosophy.

I should like to thank Mark Addis, John Brown, Michael Hallett, Dan Isaacson, Jim Lambek, Bina Ngan, Irving Rabinovitch, Annie Tong, Diana Zhang, and Grace Zhang for the inspiration and encouragement they have given me in my study of this subject. I should also like to thank Anna Sierpinska at Concordia University for the opportunity of using the penultimate draft of this book as the textbook for a one semester philosophy of mathematics course for graduate students.

W. S. Anglin

Chapter 1

Mathematics and Infinity

The key to the philosophy of mathematics is the concept of infinity. A correct treatment of infinity leads to a correct view of the whole subject, whereas an incorrect view of the infinite causes distortions.[1]

What we make of the infinite, moreover, has to do with what we make of ourselves. If human beings can, in some sense, perceive infinite totalities, then our philosophy of mathematics should reflect that fact with an openness to infinite sets or numbers. If human beings are in every respect finite, then the correct philosophy of mathematics will be one which eschews the infinite, at least as far as mathematics is seen as an activity of humans.

In what ways might humans be infinite? Either singly or collectively, human beings might be infinite in any of the following ways.
(1) They might have the capacity to use mathematical infinites to understand the physical world.
(2) There might be no upper limit on the number of facts they can know at a single time.
(3) They might be able to develop the power to calculate at an exponentially increasing rate (doing the first addition in one second, the next in half a second, the next in a quarter of a second, and so on).
(4) They might have, or be able to develop, a 'perception' of infinite sets or numbers.
(5) They might be immortal — with no upper bound on the number of calculations they can perform.

---

[1] A. George makes this point very convincingly in connection with Dummett's intuitionism. See 'How Not to Refute Realism' in the *Journal of Philosophy* 90 (1993), 53-72.

(6) They might have access to God's infinite knowledge.

(7) They might have free will in such a way that their actions cannot be explained in terms of a model which portrays humans as computers carrying out the instructions of a finite program.

(8) They might have an indefinitely expandable awareness, such that their thoughts cannot be understood in terms of any finite set of laws.

(9) They might have infinite hopes, dreams and aspirations.

If human beings are, in every respect, finite, then it is false or meaningless to describe our mathematical activity as an interaction between the human mind and the 'infinite'. However, suppose that our thinking cannot be characterised by a finite set of axioms, that it is indefinitely expandable — so that there is no mathematical proof so complex that we cannot master it. Suppose that, if only with the help of God, we can grasp infinite totalities, comparing them and ordering them. Suppose also that there is no fixed upper limit on how fast we can acquire mathematical insight. Given this very exalted view of human nature, it will not do to endorse a philosophy of mathematics which conceives it as a finite activity, dealing only with finite objects.

A philosopher's view of mathematics can be categorised as finitist or infinitist.

Augustine, for example, took a pro-infinity attitude. Since God can count through all the natural numbers, the natural numbers must exist, as a completed totality — at least as the content of an idea in God's mind. Thanks to God's grace, humans can share in God's knowledge of mathematics, so that, even here on earth, we can think in terms of the mathematical infinite. Like God, we can know that there is an infinite set of primes.[2]

Kant, on the other hand, had a finitist outlook. For Kant, God's mathematical activity (if any) is not accessible to us. Nor, for that matter, is the mathematical activity of rational creatures on other planets. What is of significance is mathematics within a human horizon, and, for Kant, that horizon is not infinitely distant. Mathematics consists of some intuitions and deductions of finite minds. Any mathematics which pretends to be infinite goes beyond the limits of human reason.[3]

Raphael's fresco *Causarum Cognitio* (at the Vatican) shows Plato and Aristotle talking philosophy. Plato is pointing upwards, as if towards the infinite, but Aristotle is stretching out his arm horizontally, well within the

---

[2] Augustine, *Gospel of John* 1.8, and *City of God* XII, 18-19.

[3] I. Kant, *Critique of Pure Reason*, B71, B182, B204, and B460.

confines of his finite universe. This difference symbolises a basic division among philosophers of mathematics. The pro-infinity attitude can be associated with Plato, while the anti-infinity attitude was vigorously argued by Aristotle himself.[4]

We shall look at some arguments in favour of the infinite but, in order to give these arguments a context, it will be useful first to review the history of the infinite in human thought.[5]

1.1 Infinity in Ancient Times

According to J. A. Benardete in his book *Infinity:*

> The whole history of mathematics might almost be written around the concept of infinity, the central theme being the various postures adopted towards finitism, both pro and con.[6]

Benardete notes that the first phase of this history begins with the Greeks, and, in this section, we look at what various Greek thinkers said about the infinite.

Anaximander (610–540)

Anaximander came from Miletus, a town in what is now south-west Turkey. His works are lost, and his thought has to be reconstructed from a single surviving quotation, and a few brief reports by later writers, such as Hippolytus (170–235), who informs us that

> Anaximander said that the principle and element of things is the infinite.[7]

---

[4]In his article 'Aristote admet-il un infini en acte et en puissance?' (*Revue Philosophique de Louvain,* 88 (1990), 487-503), A. Côté notes that some philosophers, such as J. Hintikka, have interpreted Aristotle as pro-infinity. This is not the 'classical' interpretation, and it will not be the interpretation in this book. As Côté points out, Hintikka is required to dismiss some passages in Aristotle as 'misleading' — whenever they clash with Hintikka's interpretation!

[5]The object of the following pericopes is not to give exhaustive information about what every scholar thought about what every previous scholar thought about the original thinker on the infinite. They are meant merely to show that different views on the infinite have been held, and to say a little by way of characterising or distinguishing those views.

[6]J. A. Benardete, *Infinity* (Oxford: Clarendon Press, 1964), p. 14.

[7]See M. Conche, *Anaximandre* (Paris: Presses Universitaires de France, 1991), p. 56.

Study of these brief reports suggests that, for Anaximander, there are infinitely many worlds, all made out of an infinitely extended substance, which has always existed and always will exist.[8]

However, what exactly does Anaximander mean by the 'infinite'? Is he thinking of something literally infinite, or is he merely referring to the 'indefinite' or the 'not-at-all-small'? Scholars do not agree on the answer to this question, but some, at any rate, embrace the first alternative. According to M. Conche, Anaximander's infinite is not merely a negation of the finite, but something positive and real.[9] According to L. Sweeney, Anaximander's infinite is an 'exemplification of what Aristotle calls "an actual infinite" '.[10] Sweeney adds that, for Anaximander,

> human knowledge is neither a mere inventory of individual material things nor a surface-appraisal of them. Rather, the human knower can transcend their classifications and penetrate their interiors so as to question what they really involve. *To apeiron* [the infinite] is the answer and thus, the human mind is an openness to infinity, a capacity for the infinite. [11]

The Pythagoreans (525 –)

The followers of Pythagoras listed the finite and the infinite as the first of their ten 'pairs of opposites'.[12] They preferred the finite, but they recognised the infinite. The finite was associated with rational numbers, but, as the Pythagoreans themselves proved, not all numbers are rational. The square root of 2, for example, is not the ratio of two whole numbers.[13]

As well as proving that $\sqrt{2}$ is irrational, the Pythagoreans found an infinite sequence whose limit is $\sqrt{2}$. This they did in terms of the whole number solutions of $x^2 - 2y^2 = \pm 1$. These solutions are

$$(1,\ 1),\ \ (3,\ 2),\ \ (7,\ 5),\ \ (17,\ 12),\ \ (41,\ 29),\ \ \ldots$$

If the $(n-1)$-th solution is $(x_{n-1},\ y_{n-1})$ and the n-th solution is $(x_n,\ y_n)$, then the $(n+1)$-th solution is

$$(2x_n + x_{n-1},\ 2y_n + y_{n-1})$$

---

[8] See J. Barnes, *Early Greek Philosophy* (London: Penguin, 1987), pp. 71-6.
[9] M. Conche, *Anaximandre*, ch. 2.
[10] L. Sweeney, *Infinity in the Presocratics* (The Hague: Marinus Nijhoff, 1972), p. 175.
[11] Ibid., p. 176.
[12] Aristotle, *Metaphysics* 986a15.
[13] A proof of this is found in Aristotle's *Prior Analytics* 41a23-30.

The sequence whose n-th term is x_n/y_n tends to $\sqrt{2}$, as n tends to infinity:

$$\frac{1}{1}, \ \frac{3}{2}, \ \frac{7}{5}, \ \frac{17}{12}, \ \frac{41}{29}, \ \ldots \ \rightarrow \ \sqrt{2}$$

Zeno (450 BC)

Zeno of Elea (in Italy) produced four arguments for the conclusion that there is no motion — this in support of the claim of Parmenides that being is changeless.[14]

These arguments have been studied intensely, and philosophers do not always agree about their interpretation. In what follows, we present what we feel is the best in the interpretations of Tannery, Russell, Kirk and Raven, Grünbaum, Barnes, and Sorabji.[15]

(1) The first argument concludes that motion is impossible because a moving object must first go half the distance, then half the remaining distance, and so on, forever. For example, if a point is moving from position 0 to position 1 on the number line, it first reaches position $1/2$, then position $3/4$, then position $7/8$, and so on. At the n-th stage, it is at position $1 - 1/2^n$. From the fact that there is no natural number n such that $1 - 1/2^n = 1$, Zeno concludes that the moving point never reaches position 1. Hence there is no motion — because motion from 0 to 1 is typical of any motion whatsoever.

In modern physics this argument is countered by the contention that motion can be analysed in terms of an actually infinite number of stages. In going from position 0 to position 1, the point traverses each and every one of the intervals from $1 - 1/2^n$ to $1 - 1/2^{n+1}$, for $n = 1, 2, 3, \ldots$, ad infinitum. Starting from the premiss that there is motion, modern physicists invoke the infinite to explain it.

Note that there are also finitist answers to Zeno's argument. Suppose that between position 0 and position 1 there are exactly 2,000,000 equally spaced positions which a point can occupy. The distance from 0 to 1 is thus 2,000,001 'quanta' of space.[16] The 'motion' consists of the object jumping

---

[14] Aristotle, *Physics* 239b5-240a18 and 233a21-31.

[15] See B. Russell, *Our Knowledge of the External World* (London: George Allen & Unwin, 1914), pp. 159-213; G. S. Kirk and J. E. Raven, *The Presocratic Philosophers* (Cambridge: Cambridge University Press, 1957); A. Grünbaum, *Modern Science and Zeno's Paradoxes* (Middletown: Wesleyan University Press, 1967); J. Barnes, *The Presocratic Philosophers* (London: Routledge & Kegan Paul, 1979); R. Sorabji, 'Atoms and Time Atoms' in *Infinity and Continuity in Ancient and Medieval Thought,* ed. N. Kretzmann (Ithaca: Cornell University Press, 1982), pp. 37-86; and R. Sorabji, *Time, Creation and the Continuum* (London: Duckworth, 1983), pp. 321-335.

[16] A quantum of space is a 'hodon'; a quantum of time is a 'chronon'.

discontinuously from one position to the next. Since the jumps are small, we cannot easily detect any discontinuity, but it is there all the same. To go from position 0 to position 1, the moving point first goes to a physically possible position closest to the mathematical half-way mark, say to position

$$\frac{1,000,001}{2,000,001}$$

Later it gets to position

$$\frac{1,500,001}{2,000,001}$$

At Zeno's stage 20, the moving point is at position

$$\frac{1,999,999}{2,000,001}$$

and at stage 21, it is at position

$$\frac{2,000,000}{2,000,001}$$

At stage 22, it can only jump to position 1, completing its journey in a time which is so slightly different from the time one would expect that one cannot easily detect any discrepancy.

The motion in the above, finitist account is neither uniform nor continuous. If, in agreement with most thinkers, we assume that there is uniform, continuous motion, it follows that a moving object does pass over an infinite number of points. Hence if we allow motion of the usual sort, we must allow an infinite number of points. What Zeno has, in effect, proved is that one cannot accept the existence of motion, analysing it in terms of a continuous conception of space and time, and also be a finitist.

(2) In Zeno's second argument, Achilles and the tortoise are racing along what we may take to be the positive number line.[17] Achilles starts at position 0, and the tortoise starts ahead of him, at position 1. Since Achilles runs twice as fast as the tortoise, we might expect him to catch up to the tortoise at position 2. However, when Achilles arrives at postion 1, where the tortoise started, the tortoise is already at position $1 + 1/2$; when Achilles reaches position $1 + 1/2$, the tortoise has moved on to position

$$1 + 1/2 + 1/4;$$

and so on. Whenever Achilles arrives at the tortoise's old position $2 - 1/2^n$, the tortoise arrives at position $2 - 1/2^{n+1}$ — a little ahead of Achilles.

---

[17] Actually, in our earliest account of the matter — the one due to Aristotle — the slower runner is not identified as a tortoise.

Despite the appearances which lead us to believe there is motion, Achilles will never catch up to the tortoise.

Again, Zeno is assuming that space and time are continuous (or at least dense), and that, if there is motion, there is uniform motion. Zeno also assumes that Achilles and the tortoise can never get through or complete the infinite number of stages into which Zeno, using the continuity and uniformity assumptions, has analysed their motion.

Zeno hopes that one will conclude from the paradox that the concept of motion must be discarded. The obvious reply is, again, that of modern physics: Achilles and the tortoise do get through an infinite number of stages. For modern physics, motion typically consists of the occupation of infinitely many distinct locations at infinitely many distinct instants.

(3) Zeno's third argument concludes that a flying arrow does not move because, at every instant, it is in exactly one place.

As Russell suggests, Zeno's argument is valid if one assumes that space and motion are continuous, but time is quantised — so that each interval of time can be broken up into finitely many 'atoms' of, say, $1/2^{100}$ seconds each. During each atom of time the arrow is motionless, because, if it moved from, say, 0 to 1, then there would be a time before it got to 1/2, and a time after, and so the quantum of time would be divisible into two parts. Since, during each $1/2^{100}$ second time atom, the arrow is motionless, and, since it does not jump discontinuously from one position to the next, it follows that the arrow stays at rest during its whole flight.

Of course, if one assumes that time is not quantised, but that each interval of time consists of infinitely many 'instants', each of duration 0, the fact that the arrow covers 0 distance in an instant does not imply that it covers 0 distance in an interval of such instants. As every student of calculus knows, there are cases in which $0 \times \infty$ is a positive, finite number.

(4) According to Tannery, Zeno's fourth argument presupposes the view that an interval of time is composed of finitely many time 'atoms', each lasting, say, $1/2^{100}$ seconds.[18] In this argument, there are three rows of people:

$$
\begin{array}{cccc}
A & A & A & A \\
B & B & B & B & \longrightarrow \\
\longleftarrow & C & C & C & C
\end{array}
$$

The A's are stationary, the B's move to the right at a constant rate of '1 A per atom of time', and the C's move to the left at the same speed. In $1/2^{100}$ seconds, each B moves from one A to the next A on the right, while

---

[18] Kirk and Raven support Tannery's interpretation, noting that it is 'the only way in which any sense can be made of the argument'. See page 296 in *The Presocratic Philosophers*.

each C moves from one A to the next A on the left. Given the premiss that time is quantised, this motion cannot be a steady, continuous motion. Rather, what happens is that a B, say, remains stationary, beside an A, during a half closed, half open indivisible time interval

$$[0, 1/2^{100})$$

Then, suddenly, the B disappears, and reappears beside the next A, to the right, where it spends the next indivisible time interval

$$[1/2^{100}, 2/2^{100})$$

The C's do the same thing, going the other way, so that the first B 'gets past' the first C without ever actually passing it. There is no time when they are side by side. But this contradicts our basic intuitions concerning motion.

Zeno took this contradiction to imply that there is no motion, but one could equally well take it to imply that time is not quantised, that, in other words, there are infinitely many instants in any interval of time.[19]

Each of Zeno's arguments has the following form:

Finitist assumption
Other considerations (e.g. continuity of space)
Therefore there is no motion

This is logically equivalent to the form

There is motion
Other considerations (e.g. continuity of space)
Therefore finitist assumption is wrong

Most of us accept the existence of motion, and would sooner give up finitism than embrace the static reality of Parmenides. The modern physicist, for one, is quite happy to base the analysis of motion on the mathematician's real number system, accepting the existence of infinite sets. It seems, however, that Zeno was arguing with someone who would have given up motion before giving up finitism.

---

[19]Sorabji rejects Tannery's view, and claims that Zeno's argument turns on an elementary confusion about the relativity of motion. Sorabji equates Zeno with the foolish caricature of Zeno created by Plato for the *Parmenides*. Whatever else one might say about Sorabji's interpretation, it is not charitable.

Democritus (420 BC)

Democritus of Abdera (in north-east Greece) claimed that everything is made of tiny, physical atoms. These atoms are physically (but not geometrically) indivisible. The number of atoms is infinite, and the empty space containing them is also infinite.

Democritus produced a conundrum concerning the cone. Its cross-sections are circles, and the cone can be thought of as made up of these circles. If they are finite in number, two neighbouring ones will be unequal, and the cone will be like a staircase. However, if the circles are infinite in number, two neighbouring ones will be equal, and the cone will be a cylinder.

Democritus was a determinist. He asserted that 'from infinite time back are foreordained by necessity all things that were and are and are to come'. In harmony with this, he also held that everything happens without purpose or design.[20]

Antiphon (420 BC)

Antiphon was a sophist. Noting that the area of a regular polygon is proportionate to the square on its longest diagonal, he concluded that the area of a circle is proportionate to the square on its diameter. To reach this conclusion, he assumed that a circle is a polygon with an infinite number of sides.

Plato (427–349)

Plato was a realist: he held that reality exists independently of the human mind. He was also a correspondence theorist: he held that a statement is true just in case it describes an actual state of affairs. Not surprisingly, he attacked the relativism of Protagoras, according to whom any given thing 'is to me such as it appears to me and is to you such as it appears to you'.[21]

Plato argued that the independently existing objects of reality fall into two classes: the material and the immaterial. Visible objects, such as the sun, this bed, or Diana's body, belong to the class of material things. Invisible objects, such as goodness, this circle, or Diana's soul, belong to the class of immaterial things. A drawing of a square belongs to the material realm, but the square itself, with its infinitely thin lines, belongs to the immaterial realm. Plato says of geometry students that they

---

[20] J. Barnes, *Pre-Socratic Philosophers,* vol. 2 (London: Routledge and Paul, 1979), pp. 122-4. L. Sweeney cites Democritus as an early philosopher who believed that some things are literally infinite. See L. Sweeney, *Infinity in the Presocratics,* p. 179.

[21] Plato, *Theaetetus* 152a, 160c, 169a, 193c-194b, and *Sophist* 240e-241a and 263a-d.

make use of the visible forms and talk about them, though they
are not thinking of them but of those things of which they are
a likeness, pursuing their inquiry for the sake of the square as
such and the diagonal as such, and not for the sake of the image
of it which they draw.[22]

For Plato, material things are characterised by contingency, change, un-
certainty, ignorance, and imperfection. The drawings of the square can be
erased. Their angles are not exactly right angles. Their sides are not perfectly straight. On the other hand, immaterial objects are characterised by
necessity, permanence, certainty, knowledge, and perfection. Real squares
have perfectly straight sides. We know with certainty that every square
has two exactly equal diagonals. Moreover, for Plato, real squares are not
mere abstractions or mental concepts. They are, rather, necessarily existing particulars, far more fundamental to the structure of reality than any
physical object. Finally, just as the eye sees visible objects, which exist independently of the human body, so the 'eye of the soul' intuits immaterial
objects, such as squares, which exist independently of the human soul.

At the end of Book VI of the *Republic*, Plato discusses the two classes
of things in terms of a line segment *AE* (see the Figure).

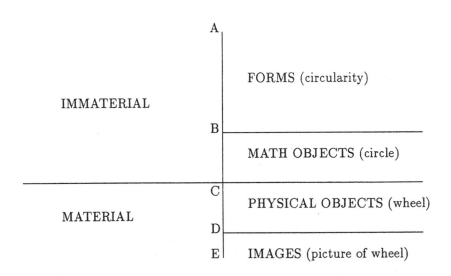

This segment is divided at a point *C* which represents the boundary between
the material realm (*CE*) and the superior, immaterial realm (*AC*). Segment

---

[22]Plato, *Republic* 510d.

AC is subdivided at B, and CE is subdivided at D. Plato tells us that

$$AC/CE = AB/BC = CD/DE$$

— from which it follows that $BC = CD$.[23] The segment DE represents pictures, reflections, or shadows of physical objects, while the segment CD represents the physical objects themselves. In the immaterial realm, segment BC represents the mathematical objects, while AB represents the *forms*. These *forms* are independently existing qualities — such as goodness, bravery, beauty, oneness, circularity, and humanity. According to Plato, a physical object is, say, circular just in case it 'participates' in the form of circularity. A picture of a wheel would be associated with segment DE, the wheel itself with segment CD. Each of two intersecting circles would be associated with segment BC, and the unique form of circularity would be associated with segment AB.[24]

For Plato, one learns about the realm of forms (AB) by examining the objects in the realm of mathematics (BC). One studies number theory 'for facilitating the conversion of the soul itself from the world of generation to essence and truth'. One studies geometry 'to facilitate the apprehension of the idea [form] of good'.[25] Having first turned his or her mind to visible geometric drawings, the student then raises it to the circles themselves, and next to the form of circularity. Finally, the student perceives the form of goodness, which illuminates all the other forms.

Plato does not say much about the infinite in his dialogues. None the less, for any natural number $n > 3$, there is a form of n-hood, and a form of n-sided polygon-hood. For example, when $n = 4$, we have the form of fourness and the form of squareness. It is not surprising, then, that Aristotle tells us that Plato holds that the infinite is 'present in' the forms, and that, for Plato, there is an infinity associated with 'the Great'.[26]

Aristotle (384–322)

Aristotle disagreed with Plato about the nature of mathematics:

> conclusions contrary to truth and to the usual views follow, if one is to suppose the objects of mathematics to exist thus as separate entities.[27]

---

[23] Ibid., 509d and 534a.
[24] See Aristotle, *Metaphysics* 987b14-17.
[25] Ibid., 525c and 526e.
[26] Aristotle, *Physics* 203a15; *Metaphysics* 987b25.
[27] Aristotle, *Metaphysics* 1077a15.

For Aristotle, the word 'two' is not a noun designating an abstract object existing separate from physical objects, but an adjective describing a physical object (e.g. that two metre ladder). The twoness of the ladder's length is not an invisible object in an invisible realm but, rather, something in the ladder itself.

Aristotle's view is human-centred. The reality of numbers has to do, not with an alien 'heaven', but with the way human beings relate to their immediate, physical surroundings. The truth about something must thus be left in abeyance if it is inaccessible to ordinary human beings. In particular, because we humans think in a finite way (says Aristotle), we cannot blithely assert the existence of infinite objects or infinite collections.

Aristotle was a staunch finitist. He rejected infinite sets and infinite lines.[28] For Aristotle, the geometer can have arbitrarily long segments, but not a line which actually 'gets to infinity'.

Aristotle gave several reasons for rejecting infinite objects or infinite collections of objects.

(1) The infinite is too big to be beautiful. In *De Poetica* 1450b–51a, Aristotle writes:

> to be beautiful, a living creature, and every whole made up of parts, must not only present a certain order in its arrangement of parts, but also be of a certain definite magnitude. Beauty is a matter of size and order, and therefore impossible ... in a creature of vast size — one, say, 1000 miles long — as in that case, instead of the object being seen all at once, the unity and wholeness of it is lost to the beholder.

(2) Infinite lines lead to contradictions in kinematics. Suppose there were an infinite straight line AB. Let C be a point not on AB, and let XCY be another infinite straight line which rotates with C as its axis, cutting AB at a variable point P. Suppose that at 3 PM, XCY is parallel to AB, and suppose that XCY rotates clockwise about C, at a constant rate of half a revolution per hour. Then XCY is parallel to AB at 4 PM, 5 PM, 6 PM, and so on — every hour on the hour. At all other times, XCY cuts AB at a point P, and, as each hour goes by, P travels the whole length of AB. However, says Aristotle, no distance is infinite if it can be traversed in a finite time. Thus AB is not infinite. Contradiction.[29]

(3) Infinite sets lead to contradictions in mathematics. If there is an infinite collection of objects, then it has a proper subset which is also infinite. For

---

[28] Aristotle, *Physics* 206b, 266b, 207; *Metaphysics* 1084a.
[29] Aristotle, *On the Heavens* 271b26–272a20.

example, the set of natural numbers contains the set of evens as a proper part, and the set of evens is infinite. However, says Aristotle, since the proper part is bounded by the whole, and less than it, the proper part is not infinite. Contradiction.[30]

(4) Aristotle also had a version of the 'Thomson Lamp Paradox'. Elaborating a bit on Aristotle, let us imagine a lamp that comes on at time $t = 1 - 1/2^n$ if n is even, but goes off at time $t = 1 - 1/2^n$ if n is odd. If, indeed, we can divide an interval of time into an actually infinite number of instants, then this lamp is theoretically possible, and, theoretically, it would turn on and off an actually infinite number of times in the time interval from $t = 0$ to $t = 1$. However, at time $t = 1$ the lamp would be neither on nor off — because the infinite is neither even nor odd. But this is impossible. Hence we cannot divide an interval of time into an actually infinite number of instants.[31]

As a replacement for the infinite, Aristotle put forward the idea of the *potentially infinite*. Imagine that Aristotle, using ruler and compass, is actually constructing the subintervals of a given segment, at a rate of one a minute. Imagine, moreover, that he will continue doing so for an indefinite period of time, so that, for any given whole number n, he will eventually construct more than n subintervals. Then, on the one hand, the set of constructed subintervals is never at any time infinite, but, on the other hand, its size is not bounded by some predetermined, fixed number. It is in this sense 'potentially infinite'.[32]

Of course, we could press Aristotle, insisting that he say something about the size of the *atemporal* set of all the subintervals that will *ever* be constructed, but, in that case, he might only reply that, like the unicorn, it is neither finite nor infinite — because it does not exist.

In this connection, it is worth noting that Aristotle does not always use the familiar two-valued logic according to which any given statement is true or else its negation is true. In discussing the future contingent proposition

A sea-battle will take place tomorrow

Aristotle says

---

[30] Aristotle, *Physics* 204a20–29 and *Metaphysics* 1066b11–17.

[31] See Aristotle, *Metaphysics* 1083b37–1084a6, and J. R. Thomson, 'Tasks and Super-Tasks', *Analysis*, 15 (1954), 1-10. In reply to Thomson, one can say that, although the lamp must be in a definite state at $t = 1$ (either on or off), it is perfectly random which state it is in. This randomness is nothing new to physics: it is, for example, perfectly random whether the Schrödinger cat is dead or alive when its box is opened.

[32] Aristotle, *Physics* 206a18–26 and *Metaphysics* 1048b10–18.

it is not necessary that of every affirmation and opposite negation one should be true and the other false.[33]

For Aristotle it is not necessarily illogical to deny that

A sea-battle will take place tomorrow

and also, simultaneously, to deny that

A sea-battle will not take place tomorrow.

Similarly, for Aristotle, it is not necessarily illogical to deny that

There are infinitely many subintervals

and also to deny that

There are finitely many subintervals.

For Aristotle, the complete set of subintervals of a given segment — like
the merely possible sea-battle — is something which does not exist yet,
and hence we need not feel that it has to be, in every respect, one way
(e.g. finite) rather than the other (e.g. infinite). If we are confident that
standard, two-valued logic is right, then we may dismiss this as a dodge.
However, it should be noted that there is a school of philosophers which
use a coherent 'intuitionistic logic' which allows the move Aristotle needs
to make. As we note below, the 'intuitionists' are the twentieth century
custodians of the anti-infinity attitude in the philosophy of mathematics.

Euclid (300 BC)

Aristotle's finitism seems to have had an effect on Euclid's *Elements*. For
Euclid, one of the first professors at the University of Alexandria, geometry
does not start with an infinite set of lines. Rather the lines have to be
constructed, one by one. Furthermore, these lines do not have infinite
length: they are finite segments. It is true that they can always be extended,
but only by finite lengths, and only a finite number of times. Nor are these
lines infinite sets of points, but each line is a single entity in its own right.
For Euclid, parallels AB and CD are not infinitely long lines which 'meet at
infinity'. Rather, they are finite segments such that one cannot construct
a point P which is on AB (or AB produced), and also on CD (or CD
produced).

In the *Elements,* the 'domain of discourse' is not a static, infinite set,
but, on the contrary, a growing finite set. It is 'potentially infinite', but it

---

[33] Aristotle, *De Interpretatione* 9. See also *Physics* 202b30-35.

is never, at any time, infinite. Points and lines are added as necessary, but only a finite number of them. It is true that Euclid sometimes talks about taking a point 'at random' — suggesting, perhaps, a reference to a set of points given prior to any construction — but it is clear from Euclid's overall approach that what this actually means is it that it does not matter which particular point one *constructs*. For example, to take a point 'at random' on the segment AB is to use the straightedge and compass to construct some point or other on AB, a point that is arbitrary, yes, but not so arbitrary that it is not constructible.

An axiomatisation of the *Elements* that was historically accurate would not presuppose a fixed, infinite set of points. It would not begin with Platonic existence claims. It would not, for example, have the following version of the Parallel Postulate:

> Given distinct, noncollinear points A, B, C, and points D and E with $D \neq E$, there is a point F such that F is on line DE, and also either (1) F is on line AB or (2) F is on line AC.[34]

Rather, it would have the following 'basic construction':

> Given distinct, noncollinear points A, B, C, and points D and E with $D \neq E$, *one can construct* a point F such that F is on line DE, and also either (1) F is on line AB or (2) F is on line AC.

The 'meeting point' F is not, for Euclid, some already existing member of, say, $\mathbf{R} \times \mathbf{R}$, but, rather, a point which may have to be added to the finite set of points already constructed.

Archimedes (287-212)

As we know from his book *The Method*, Archimedes of Syracuse (in Sicily) accepted the infinite, using many of the techniques found in calculus. For example, he regarded areas as infinite sums of line segments.

To establish the formula for the area of a circle, Archimedes assumed, in effect, that

$$\lim_{n \to \infty} 2^n \sin \frac{180°}{2^n} = \lim_{n \to \infty} 2^n \tan \frac{180°}{2^n}$$

— this limit being the number we call π. The fact that he could not construct a segment of length π using only a finite number of straightedge and compass operations did not stop him. Archimedes accepted the infinite in a way Euclid did not.

---

[34] It is the inclusive 'or' that is meant here.

Nicomachus (100 AD)

Nicomachus of Gerasa (near Jerusalem) was a Neo-Pythagorean. He wrote a number theory book, the *Introductio arithmeticae,* in which he notes that if the odd numbers are arranged in an equilateral triangle (see the Figure), then the sum of the odd numbers in the n-th row of the triangle is n^3. For example, the sum of the numbers in the third row is $7 + 9 + 11 = 27 = 3^3$.

$$1$$
$$3 \qquad 5$$
$$7 \qquad 9 \qquad 11$$
$$13 \qquad 15 \qquad 17 \qquad 19$$

The *Introductio arithmeticae* also contains some Pythagorean philosophy. Nicomachus was one of the first thinkers to locate the natural numbers in the Divine Mind.[35] Nicomachus compared arithmetic with the other branches of mathematics, writing:

> it existed before all the others in the mind of the creating God like some universal and examplary plan.[36]

In the nineteenth century, Nicomachus's view was held by Edward Everett:

> in pure mathematics we contemplate absolute truths, which existed in the divine mind before the morning stars sang together, and which will continue to exist there, when the last of their radiant host shall have fallen from heaven.[37]

In the twentieth century, Nicomachus's view was held by Ramanujan (1887–1920):

> an equation for me has no meaning unless it expresses a thought of God.[38]

In any case, from Nicomachus on, we must distinguish three sorts of pro-infinity views:
(1) God exists and is infinite; however, there are no infinites in mathematics;
(2) God exists and is infinite; also, there are infinites in mathematics;
(3) God does not exist; none the less there are infinites in mathematics.

---

[35] F. Copleston, *A History of Western Philosophy,* vol. 1 (New York: Doubleday, 1985), p. 447.

[36] See Victor J. Katz, *A History of Mathematics* (New York: HarperCollins, 1993), p. 158.

[37] *Orations and Speeches* (Boston, 1870), Vol. 3, p. 514.

[38] See page 7 in *The Man Who Knew Infinity* by R. Kanigel (New York: Charles Scribner's Sons, 1991).

View (2) is often elaborated by the view that God is the cause or reason for the infinites in mathematics. The set of natural numbers is infinite because God conceives of an infinite number of natural numbers. The Axiom of Choice is true because God conceives the choice sets.

1.2 Infinity in the Middle Ages

The ancient Greeks were unable to settle the problem of the infinite. The Neo-Platonists were for it, at least in the case of God being infinite, but Boethius (475–524), for example, was against it, at least in the matter of mathematics:

> nothing which is infinite is able to be assembled by a science or to be comprehended by the mind.[39]

Moving into the Medieval period, we find that thought on the infinite is influenced by theology. Thanks partly to Aristotle, some philosophers associated the infinite with incomplete magnitude. Since God is neither incomplete nor a magnitude, these philosophers did not want to say that God is infinite — although they did agree that the magnitude of his power is infinite. Other philosophers saw the infinite in a more positive, less mathematical light, and were willing to say that God himself is, in some sense, infinite. For example, Gregory of Nyssa (331–395), Augustine (354–430), and John Damascene (750) all taught that God is infinite, and, thanks to Aquinas (1225–1275), this has become a *de facto* dogma.[40]

When we get to the fourteenth century, we find that mathematicians are doing work which presupposes and explores the actually infinite in mathematics. There is, moreover, a theological flavour to their work which suggests that it is pro-infinity theology that motivates them. For example, Gregory of Rimini (1300–1358) considers an infinite series in order to show that God can create an actually infinite stone. Jacques Sesiano believes — rightly I think — that 'theological considerations led medieval philosophers to reconsider Aristotle's statements about infinity'.[41]

---

[39] Boethius, *Boethian Number Theory*, trans. M. Masi (Amsterdam: Rodopi, 1983), p. 73.

[40] For a detailed account of the history of applying the adjective 'infinite' to God, see L. Sweeney, *Divine Infinity in Greek and Medieval Thought* (New York: Peter Lang, 1992).

[41] Jacques Sesiano, 'On an algorithm for the approximation of surds from a Provençal treatise' in *Mathematics from Manuscript to Print*, ed. C. Hay (Oxford: Clarendon Press, 1988), p. 40.

Augustine (354-430)

Augustine of Hippo (in present-day Algeria) was a Christian with great respect for Plato, and he shared Plato's enthusiasm for mathematics. In the *City of God* Augustine writes:

> Six is a number perfect in itself, and not because God created all things in six days; rather, the converse is true. God created all things in six days because this number is perfect, and it would have been perfect even if the work of six days did not exist.[42]

According to Augustine, the word 'infinite' applies to God's power and also to his knowledge. The divine mind is

> infinitely capacious and utterly immutable, a mind that can count uncountable things without passing from one to the next.[43]

For Augustine, the natural numbers form an infinite totality which is understood, if not by us, certainly by God. Although we do not know if there are infinitely many perfect numbers, God does know this, and he knows this by means of a direct, simultaneous inspection of each and every natural number.

Augustine takes pains to dispel the notion that God might not be interested in numbers:

> the philosophers, those at least who respect Plato's idea that God designed the cosmos on the principle of numbers, will hardly dare to despise numbers, in the sense of pretending that they are no part of the knowledge of God. And as for us [Christians], our Scripture thus addresses God: 'Thou hast ordered all things in measure, and number, and weight,' and ... in the Gospel, our Saviour [Jesus] declares that 'the very hairs of your head are all numbered.'[44]

Bhaskara (1114-1185)

Bhaskara, an Indian mathematician, was the first person to find a positive integer solution to the Diophantine equation $x^2 - 61y^2 = 1$. Like Augustine,

---

[42] Augustine, *City of God* XI 30. A whole number is *perfect* if it is the sum of its proper divisors: for example, 6 is perfect because 6 has proper divisors 1, 2, and 3, and, moreover, $1 + 2 + 3 = 6$. Euclid had given a proof of the fact that if p is a prime which is one less than a power of 2, then $p(p+1)/2$ is perfect: for example, 3 is a prime which is 1 less than 2^2 and $3(3+1)/2 = 6$ is perfect.

[43] Ibid., XII 18.

[44] Ibid. The Bible quotations are *Wisdom* 11:20 and *Matthew* 10:30.

he was moved at least partly by theological considerations to accept an actual infinite:

> Quotient the fraction $\frac{3}{0}$. This fraction, of which the denominator is zero, is termed an infinite quantity. In this quantity ... there is no alteration, though many be inserted or extracted; as no change takes place in the infinite and immutable God, at the period of the destruction or creation of worlds, though numerous orders of beings are absorbed or put forth.[45]

Aquinas (1225–1275)

There is a passage in his *Summa Theologiae* which suggests that Thomas Aquinas held that there is no infinite in mathematics:

> no set of things can actually be inherently unlimited, nor can it [just] happen to be unlimited.[46]

From this, and similar passages, M. Hallett has concluded that Aquinas should be classified with those who believe in an infinite God, but deny that there is any mathematical infinite.[47] However, other passages — as well as the logic of Aquinas's overall position — indicate that he should be counted with Augustine. For example, in Part I, Article 14, Question 12 of the *Summa*, Aquinas agrees with Augustine that God knows all possible things (that is, all the entities he either does or might create). Moreover, he knows all the future thoughts of immortal beings (such as ourselves). The collection of possible things, and the collection of future thoughts are both infinite, but, *pace* Aristotle, God none the less has a complete knowledge of them. Aquinas follows Aristotle whenever possible, but, in connection with God's knowledge, Aquinas is compelled to agree with Augustine that, distinct from the infinite who is God himself, there are mathematical infinites, such as infinite collections.[48]

Gregory of Rimini (1300-1358)

Gregory of Rimini (in Italy) maintained that God could create an actually infinite stone. Gregory explained that God could do this by creating equal

---

[45] W. S. Anglin, *Mathematics: A Concise History and Philosophy* (New York: Springer-Verlag, 1994), p. 116.

[46] Aquinas, *Summa Theologiae* 1a7, Q4.

[47] M. Hallett, *Cantorian Set Theory and Limitation of Size* (Oxford: Clarendon Press, 1984), pp. 13 and 22-3.

[48] See J. J. MacIntosh's article 'St. Thomas and the Traversal of the Infinite' (*American Catholic Philosophical Quarterly*, 68 (1994), 157-177).

sized bits of the stone at each of the times $t = 0$, $1/2$, $3/4$, $7/8$, ..., and so on.[49] Gregory also asserted that God could create an actually infinite number of angels 'within an hour'.[50]

Albert of Saxony (1350 AD)

Albert of Saxony showed that one can sometimes take a proper subset of an infinite set, and re-arrange its elements, so that it shows itself to be just as big and unbounded as the infinite set of which it is a proper part. Specifically, he noted that if one has an infinitely long beam of wood, with equal width and depth, one can saw it up into equal sized cubic blocks with which one can fill the whole of what we call Euclidean 3-space. (Surround the first block with $3^3 - 1$ more blocks, making a cube of side 3; then surround that cube with $5^3 - 3^3$ more blocks, making a cube of side 5; and so on.) In modern terminology, what Albert proved is that there is a one-to-one correspondence between the set of triples $(n, 1, 1)$, with n a positive integer, and the set of triples (a, b, c), with a, b, and c any integers.

    Aristotle had objected to the infinite on the grounds that an infinite set would have a proper subset that was also infinite, even though it was only a fraction of the size of the original set, and thus not really infinite, at least not in comparison with the original set. What Albert of Saxony showed is that this proper subset is, in one way, just as infinite as the original set.[51]

Oresme (1350 AD)

Bishop Nicole Oresme was the first mathematician to prove the divergence of the harmonic series

$$\frac{1}{1} + \frac{1}{2} + \frac{1}{3} + \frac{1}{4} + \cdots$$

He also found the (finite) sum of the infinite series

$$\frac{1}{2} + \frac{2}{4} + \frac{3}{8} + \frac{4}{16} + \ldots + \frac{n}{2^n} + \cdots$$

One of Oresme's examples concerns a solid which covers the whole plane, but has only a finite volume. It consists of a cylinder of radius 1 and height $1/(2\pi)$, surrounded by a cylindrical ring of width 1 and height

$$\frac{1}{2^2\pi(2^2 - 1^2)}$$

---

[49] A. W. Moore, *The Infinite* (London: Routledge, 1991), p. 53.

[50] G. Leff, *Gregory of Rimini* (Manchester: Manchester University Press, 1961), pp. 124-127. Leff gives details about the location of our oldest copies of Gregory's books.

[51] Jacques Sesiano, p. 41. Sesiano believes that Albert understood his conclusion to mean that two infinites are not comparable.

and this, in turn, surrounded by a cylindrical ring of width 1 and height

$$\frac{1}{2^3\pi(3^2 - 2^2)}$$

and so on. The volume of the infinite solid works out to be 1.[52]

[52]N. Oresme, *Nicole Oresme and the Medieval Geometry of Qualities and Motions,* trans. M. Clagett (Madison: University of Wisconsin Press, 1968), pp. 427-431.

Chapter 2

Mathematics and Infinity II

In this chapter we continue, and conclude, our account of the notion of infinity in the history of ideas.

2.1 Infinity at the Beginning of the Modern Era

The God-centred philosophy of the Middle Ages yielded to a more human-centred philosophy in the modern period. Schools such as empiricism rejected the metaphysical munificence associated with divine revelation, and advocated a parsimony based on the limits of ordinary human perceptions. As a result, the infinite once more came under suspicion, a suspicion which eventually led some mathematicians, in the nineteenth century, to try to put all of mathematics on a finitist footing. From the time of Descartes to the time of Cauchy, mathematicians worked with infinites (and the corresponding 'infinitesimals'), but they were uneasily aware that many philosophers were sceptical. During the Middle Ages, philosophers had tried to convince mathematicians that the infinite was respectable, but during the beginning of the modern period, philosophers tried to convince them that it was not.

Descartes (1596–1650)

Boethius had claimed that nothing infinite is able 'to be comprehended by the mind', but René Descartes believed that we possess a 'clear and distinct idea' of the infinite. In *Meditation* III, he asserts that our concept of the infinite is logically and even epistemologically prior to our idea of the finite:

> I clearly perceive that there is more reality in the infinite sub-
> stance than in the finite, and therefore that in some way I pos-
> sess the perception (notion) of the infinite before that of the
> finite.[1]

In spite of his pro-infinity philosophy, Descartes has mixed views about the infinite in mathematics. Sometimes he is against it. P. Mancosu notes that:

> His rejection of the mechanical curves [such as the quadratrix]
> is grounded in the idea that their construction involves us in
> infinite processes of approximation which cannot be exact (ge-
> ometrical). His method of tangents also exemplifies his careful
> avoidance of infinitesimal arguments.[2]

There are other times, however, when Descartes accepts the infinite in mathematics. For example, in *The Geometry*, he states that a 'locus', such as a straight line or circle, is an infinite set of points.[3]

Pascal (1623–1662)

Blaise Pascal accepted the infinite in both religion and mathematics. It was he who described humans as finite creatures located between the infinitely large and the infinitely small. He exclaims: 'the everlasting silence of those infinite expanses frightens me!'[4]

In mathematics, Pascal made use of the new projective geometry of Girard Desargues (1591–1661) with its 'points at infinity', and he gave us the first clear statement of the Principle of Mathematical Induction, this enabling one to prove formulas covering, as Pascal says, 'an infinity of cases'.[5]

---

[1] R. Descartes, *Meditations,* in *The Rationalists* (New York: Doubleday, 1960), p. 137. See also *The Philosophical Works of Descartes,* vol. 1, trans. E. S. Haldane and G. R. T. Ross (New York: Dover, 1953), pp. 229-30.

[2] P. Mancosu, 'Descartes's *Géométrie*', in *Revolutions in Mathematics,* ed. D. Gillies (Oxford: Clarendon Press, 1992), p. 102.

[3] R. Descartes, *The Geometry,* trans. D. E. Smith (New York: Dover, 1954), pp. 23 and 35.

[4] B. Pascal, *Oeuvres complètes* (Tours: Gallimard, 1954), p. 1113.

[5] Ibid., p. 103, 584-92.

Hobbes (1588-1679)

Thomas Hobbes promoted the empiricist notion that human beings are finite and therefore cannot conceive the infinite:

> Whatever we imagine is *finite*. Therefore there is no idea, or conception of anything we call *infinite*. No man can have in his mind an image of infinite magnitude; nor conceive infinite swiftness, infinite time, or infinite forces, or infinite power.[6]

Of course, a person might mistakenly believe they could conceive the infinite, but that would be on a par with mistakenly believing one could conceive a purple geometric line.

Note that, for someone, like Hobbes, with a human-centred metaphysics, Hobbes's denial is not just a denial of our concept of the infinite, but a denial of the infinite itself.

In 1641, Evangelista Torricelli (1608–1647) proved that an infinitely long solid could have a finite volume. His example was the trumpet-shaped solid of revolution generated by the curve $y = 1/x$, with $x \geq 1$. Torricelli's result shocked many seventeenth century thinkers, including Hobbes, whose own reaction was to deny that the solid was infinitely long.[7] P. Mancosu states that Hobbes's

> mathematical analysis of Torricelli's cubature is finally ill conceived and fails to do justice to the striking originality of the theorem.[8]

Locke (1632–1704)

John Locke's *Essay* contains a section called 'Of the Idea of Infinity' in which Locke adopts a position redolent of Aristotle:

> to have actually in the mind the idea of a space infinite, is to suppose the mind already passed over, and actually to have a view of *all* those repeated ideas of [finite] space which an *endless* repetition can never totally represent to it; which carries in it a plain contradiction.

---

[6] T. Hobbes, *The English Works of Thomas Hobbes,* III (London: John Bohn, 1839), p. 17.

[7] See the chapter on paradoxes in P. Mancosu, *Philosophy of Mathematics and Mathematical Practice in the Seveenth Century* (New York: Oxford University Press, 1996). Apparently, seventeenth century thinkers were unaware of the fact that Torricelli's theorem had been proved three hundred years previously, by Bishop Oresme.

[8] Ibid., p. 149.

Shifting his attention from geometry to arithmetic, Locke adds:

> there is nothing yet more evident than the absurdity of the
> actual idea of an infinite number.

Moreover,

> if a man had a positive idea of infinite, either duration or space,
> he could add two infinites together; nay, make one infinite in-
> finitely bigger than another — absurdities too gross to be confuted.[9]

In the twentieth century, of course, set theorists quite routinely 'add two infinites together; nay, make one infinite infinitely bigger than another'.

Leibniz (1646–1716)

In *New Essays,* Gottfried Wilhelm Leibniz agrees with Locke:

> there is no infinite number, neither line nor other infinite quantity.[10]

However, in some of his letters, Leibniz qualifies this position. In a letter to Simon Foucher, written in 1693, Leibniz endorses a 'Principle of Plentitude', according to which

> the smallest particle should be considered as a world full of an
> infinity of creatures.[11]

In a letter to P. des Bosses, dated March 1706, Leibniz asserts that all possible natural numbers exist in the mind of God.[12]

Leibniz was one of the discoverers of the calculus. To find the slope of the tangent of, say, the parabola $y = x^2$ at the point (x, x^2), he would reason as follows. Let dx be an infinitely small, or 'infinitesimal', increment of x. The corresponding increment of y is then

$$dy = (x + dx)^2 - x^2 = 2x(dx) + (dx)^2$$

This represents the rise corresponding to the run of dx, and hence the slope of the tangent at x is

$$\frac{\text{rise}}{\text{run}} = \frac{dy}{dx} = 2x + dx$$

Hence, now equating the dx to 0, the tangent at x has slope $2x$.

For Leibniz, infinitesimals were, not 'real' numbers, but useful 'fictions'.

---

[9] J. Locke, *Essay* I, ed. A. C. Fraser (Oxford: Clarendon Press, 1894), II 17 (especially pp. 281 and 292).

[10] G. Leibniz, *New Essays* (Cambridge: Cambridge University Press, 1981), II 17.

[11] Benardete, *Infinity,* p. 19.

[12] G. W. Leibniz, *Philosophische Schriften* V (Frankfurt am Main: Insel Verlag, 1990), p. 230.

Berkeley (1685–1753)

George Berkeley was willing to call God 'infinite', but he did not think the infinite should be allowed in mathematics. In *A Treastise concerning the Principles of Human Knowledge* (I 131), he writes:

> there is in effect no such thing as parts infinitely small, or an infinite number of parts contained in any finite quantity [e.g. a line segment].

Berkeley is famous for his attack on infinitesimals. In *The analyst*, written in the year in which Berkeley became an Anglican bishop, he mocks mathematicians who are willing to accept shoddy reasoning about the infinitely small, but are not willing to accept Christian teachings:

> he who can digest a second or third fluxion, a second or third difference, need not, methinks, be squeamish about any point in divinity.

One of Berkeley's arguments was this: either the infinitesimal dx is 0 or it is not. If it is 0, you cannot divide by it (as Leibniz had). If it is not 0, then you cannot ignore it in a sum (as Leibniz had). Either way, there is a problem.

Hume (1711–1776)

David Hume denied our ability to conceive the infinite, and he also denied the existence of anything infinite.

> The capacity of the mind is not infinite; consequently no idea of extension or duration consists of an infinite number of parts.[13]

Furthermore, we ought to

> regard all the mathematical arguments for infinite divisibility [of line segments] as utterly sophistical.[14]

One of Hume's own arguments is this: if line segments are made up of infinitely many points, then all line segments are equal in length (since every line segment has the 'same number' of points in it). However, not all line segments are equal in length. Hence line segments are not made up of infinitely many points.

---

[13] D. Hume, *A Treatise of Human Nature*, ed. L. A. Selby-Bigge (Oxford: Clarendon Press, 1896), p. 39.
[14] Ibid., p. 53.

Needless to say, Hume is confusing cardinality and measure. He is also disregarding two possibilities that were accepted by mathematicians of his day: (1) any point has length 0, and (2) numbers of the form

$$0 \times \infty$$

can take many different finite values.

In his article, 'Achievements and Fallacies in Hume's Account of Infinite Divisibility', J. Franklin argues that Hume's troubles with the infinite are due to his 'a priori atomism' — his view that it is impossible that a 'point' not have some nonzero size (the same size for every point). Franklin notes that

> almost all commentators, even ones who usually admire Hume, have judged his conclusion on infinite divisibility to be false, and his reasons so hopelessly confused as to be of no interest.[15]

Kant (1724–1804)

Immanuel Kant was not an empiricist, but he retained the same human-centred outlook. Like Aristotle, Kant thought in terms of a potential infinity. In the *Critique of Pure Reason*, Kant considers the possibility that there is no first human being, that each of us has an unlimited number of ancestors. Kant claims that, in such a case, the sequence of ancestors is neither finite nor infinite 'for it is nothing in itself'.[16] Kant says a similar thing about the possibility that the universe has no finite age. If that is the case, 'the magnitude [duration] of the world can be taken neither as finite nor as infinite'.[17] Kant adds:

> since the world is not given to me, in its totality, through any intuition, neither is its magnitude given me.[18]

Aristotle, recall, had suggested that, where p is the proposition

> There will be a sea-battle tomorrow

neither p nor not-p is true — since the truth of either of these propositions would depend on objective events which do not yet exist. Kant applies this suggestion in the case of infinite sequences. Since they are neither 'given' to the human mind, nor required by the human mind in its basic

---

[15] J. Franklin, 'Achievements and Fallacies in Hume's Account of Infinite Divisibility', *Hume Studies,* 20 (1994), 86.

[16] I. Kant, *Critique of Pure Reason,* B542.

[17] Ibid., B546.

[18] Ibid., B547.

thinking, they are not 'there' so that we could correctly describe them as, say, infinite. For Kant, an unlimited sequence cannot be called finite, because it 'grows' bigger without limit, and it cannot be called infinite either, since its 'growing' is never completed. Kant writes:

> I cannot say, therefore, that the world is *infinite* in space or as regards past time. Any such concept of magnitude, as being that of a given infinitude, is empirically impossible, and therefore, in reference to the world as an object of the senses, also absolutely impossible. Nor can I say that the regress from a given perception to all that limits it in a series, whether in space or in past time, proceeds to *infinity;* that would be to presuppose that the world has infinite magnitude. I also cannot say that the regress is *finite;* an absolute limit is likewise empirically impossible.[19]

Twentieth century 'intuitionists' use a logic in which the law of the excluded third (viz. p or not-p) is rejected. Within this logic, one may deny that the world is either finite or infinite. However, it turns out to be very limiting to restrict oneself to this logic, and also very limiting to deny that there are infinites in mathematics. For example, one cannot accept some of the basic theorems in calculus. Nor is this surprising. If one starts with the empirical dogma that truth is circumscribed by finite perceptions, one will naturally enough end up with fewer truths.

2.2 Infinity in the Nineteenth Century

Empiricism influenced many mathematicians in the nineteenth century. Carl Friedrich Gauss (1777–1855) asserted that 'infinity is merely a façon de parler', and, in his *Disquisitiones Arithmeticae,* he took pains to give a solution of the Diophantine equation

$$x^2 - Ry^2 = C$$

which, unlike the previous solution given by J. L. Lagrange (1736–1813), made no reference to limits of infinite sequences. A. L. Cauchy (1789–1857) and K. Weierstrass (1815–1897) devoted much energy to reducing the concepts of limit, convergence, and so on, to 'epsilon-delta' definitions, these containing no mention of the infinite. L. Kronecker (1823–1891) rejected all infinite sets.[20] There were, however, some nineteenth century mathematicians who were open to the infinite.

---

[19] Ibid., B548.

[20] D. M. Burton, *The History of Mathematics,* 2nd edn. (Dubuque: William C. Brown, 1991), p. 592.

Bolzano (1781–1848)

Bernard Bolzano wrote a book called *Paradoxes of the Infinite* in which he discussed, for example, the fact that a geometric figure with a finite interior can have an infinite boundary.[21]

Addressing the 'paradox' of an infinite set having infinite parts, Bolzano distinguishes set equality and set equipotence. Two sets, A and B, are *equal* ($A = B$) just in case they have the same members. Two sets, A and B, are *equipotent* ($\#A = \#B$) just in case there is a one-to-one mapping from A to B. In the case of finite sets, we have what might be called *Aristotle's Principle:*

$$\text{If } A \subseteq B \text{ and } \#A = \#B \text{ then } A = B$$

Bolzano notes that the fact that this breaks down for infinite sets does not imply that there are no infinite sets; it just means that one can use Aristotle's Principle to distinguish infinite from finite sets. Unfortunately, Bolzano went further than this, and asserted that

$$\text{If } A \subseteq B, \text{ and both } A \text{ and } B \text{ are infinite then } \#A = \#B$$

— a principle which was refuted by Cantor.[22]

Dedekind (1831–1916)

Richard Dedekind embraced the actual infinite as early as 1856, and he used it, in 1858, in his theory of real numbers.[23] In 1888, Dedekind gave a pro-infinity argument, which we shall present below.

Cantor (1845–1918)

Georg Cantor produced a clear and complete theory of the infinite which answered objections that had previously been raised by anti-infinity philosophers, and which is now regarded as the cornerstone of mathematics. Thanks to Cantor we have a new and deeper understanding of the real numbers, and of all the branches of mathematics which presuppose them. Because of the anti-infinity prejudices of Kronecker and others, Cantor never obtained a position at a first-rate university, but history has proved

---

[21] B. Bolzano, *Paradoxes of the Infinite,* trans. Fr. Prihonsky (London: Routledge and Kegan Paul, 1950), p. 140. One such figure is Toricelli's solid, generated by rotating the curve $y = 1/x$, with $x \geq 1$, about the x-axis: its volume is finite, although its surface area is infinite. Another example is the fractal called the 'von Koch snowflake'.

[22] Ibid., p. 96.

[23] J. Ferreirós, 'Traditional Logic and the Early History of Sets', *Archive for History of Exact Sciences,* 50 (1996), 53.

him to be one of the most original and most profound mathematicians of all time. The opening sentence in M. Hallett's *Cantorian Set Theory and Limitation of Size* is no exaggeration:

> Cantor was the founder of the mathematical theory of the infinite, and so one might with justice call him the founder of modern mathematics.[24]

Cantor believed in an infinite God, and also in infinite sets. For Cantor, the latter belief was justified by the former:

> Since God is of the highest perfection one can conclude that it is possible for Him to create a *transfinitum ordinatum* [realm of the infinite]. Therefore, in virtue of His pure goodness and majesty we can conclude that there actually is a created *transfinitum*.[25]

Here Cantor invoked Augustine's Principle of Plenitude, which states that a good God creates every kind of good thing.[26]

Cantor's belief in God also helped lead him to the view that not every collection of objects is itself eligible to be a member of a collection. Cantor held this partly because he believed that the collection consisting of 'everything' is divine, and in such a way that it is over-qualified to be a member in what would be some 'higher' collection. Some of the early workers in set theory, such as Bertrand Russell (1872–1970), originally thought that no such restriction was necessary. They espoused the 'naive' view that any collection was eligible for membership in any collection. A collection could even be a member of itself. This left them open to a contradiction discovered first by Ernst Zermelo (1871–1953), and then, independently, by Russell, which is now called the *Russell Paradox:*

> Let C be the collection such that, for any collection X,
> X is a member of C if and only if X is not a member of X.
> Then C is a member of C if and only if C is not a member of C.

---

[24] M. Hallett, *Cantorian Set Theory and Limitation of Size,* p. 1.

[25] Ibid., p. 23.

[26] The Principle of Plenitude implies that if, say, elephants are good, then, since it is necessary that God is good, it is also necessary that God create elephants. Hence it is necessary that elephants exist. Some thinkers (such as Aquinas) associate necessary existence with divinity, and, for these thinkers, it would follow that the elephants were divine. Hence, for these thinkers, the Principle of Plenitude is redolent of a kind of pantheism. Anne Newstead has told me that it may have been because he did not want to be connected, in any way, with pantheism, that Cantor had second thoughts about endorsing the Principle of Plenitude. See J. W. Dauben, *Georg Cantor* (Cambridge, Mass.: Harvard University Press, 1979), pp. 145-6.

The now standard 'von Neumann-Bernays-Gödel set theory' distinguishes, right from the start, those collections, called *proper classes,* which are not eligible for membership in collections, from those collections, called *sets,* which are. Examples of proper classes are the collection of all sets, and also the Russell collection defined above. Examples of sets are the collection of natural numbers, and the collection of Euclidean points. Cantor's theological intuitions contributed to saving him from the contradictions of 'naive set theory'.[27]

A similar thing happened in the case of the Well Ordering Principle, an axiom equivalent to the Axiom of Choice. Cantor adopted it because he believed there is a God who can arrange the elements of any set so that they are well-ordered. As it was discovered later, the Well Ordering Principle plays a key role in many branches of mathematics. Cantor's faith in God guided him in the right direction.[28]

One of Cantor's striking results is that there is an infinite hierarchy of distinct infinites, each infinitely greater than those below it. The medievals had noted that the number of points in a large circle is the same as that in a small concentric circle, in the sense that each radius of the large circle passes through exactly one point of each circle. Similar observations had led Bolzano to the conclusion that any two infinite sets are equipotent. In 1873 Cantor discovered that this is wrong. One of his proofs goes as follows.

Let A be an infinite set (that is, one containing infinitely many members). Let $P(A)$ be the set of subcollections of A. Suppose that A and $P(A)$ are linked by a one-to-one correspondence $f : A \rightarrow P(A)$. Let S be the collection of members x of A such that x is not a member of $f(x)$. Then S is a member of $P(A)$, and, on the assumption, there is some member y of A such that $S = f(y)$.

If $y \in S = f(y)$ then, by the defining property of S, y is not a member of $f(y)$. However, if y is not a member of $f(y) = S$, then, by the definition of S, it is a member of S. Contradiction. Hence A and $P(A)$ are not linked by a one-to-one correspondence.

---

[27]Hallett, *Cantorian Set Theory and Limitation of Size,* Chapter 1, and p. 168. In 1988, Dauben published an article in which he claimed that Cantor may have been saved from the contradictions of naive set theory, not by his theology, but by an awareness of those contradictions. According to Dauben, Cantor may have discovered at least one of them as early as 1882. The more common view, expressed, for example, by the earlier Dauben (on pages 226 and 245 of his 1979 book *Georg Cantor*) is that the contradictions of naive set theory were not discovered until about 1895. Indeed, if the later Dauben is right, why did Cantor keep his knowledge of the contradictions hidden for over 10 years? See Dauben's 'Cantorian Set Theory and Limitations of Size', *British Journal for Philosophy of Science,* 39 (1988), 542.

[28]Hallett, *Cantorian Set Theory and Limitation of Size,* pp. 156-7. See also S. Lavine, *Understanding the Infinite* (Cambridge: Harvard University Press, 1994), pp. 55, 117, 289.

Since, for every member x of A, $P(A)$ has $\{x\}$ as a member, there is a 'copy' of A which is a subset of $P(A)$. Hence A is smaller than $P(A)$, and we can write $A < P(A)$. Similarly, $P(A) < P(P(A))$. Indeed, we have an infinite hierarchy of infinite sets, each more infinite than the previous ones:

$$A < P(A) < P(P(A)) < P(P(P(A))) < P(P(P(P(A)))) < \cdots$$

Results like this incurred the scorn of Kronecker.[29]

Cantor raised the following question. If A is the set of positive integers, we know that $A < P(A)$, but is there some set B such that $A < B$ and $B < P(A)$? Cantor conjectured that the answer is 'no', and it is this conjecture which is called the *Continuum Hypothesis*. It has been proved, by Kurt Gödel and Paul Cohen, that neither the Continuum Hypothesis nor its negation follows (in first order logic) from the 'basic' axioms of set theory, and no one has yet been able to produce a not-so-basic axiom which would yield a convincing answer to Cantor's question.

Note, however, that this does not mean that the Continuum Hypothesis is the Parallel Postulate of Set Theory. For whereas the Parallel Postulate and its negation both show up in different interpretations of 'Euclid's other axioms' — even when those axioms are given in terms of 'second order logic' — a similar thing does not happen with the Continuum Hypothesis. In 1930 Zermelo showed that the standard set theory axioms — given in terms of second order logic — have only one kind of interpretation as far as the Continuum Hypothesis is concerned. Either the Continuum Hypothesis is 'true in all interpretations' or else its negation is 'true in all interpretations'. Hence, although we do not yet know which of the alternatives holds, we do know that there is a real 'fact of the matter'. Cantor was addressing a genuine issue.[30]

2.3 Infinity in the Twentieth Century

The vast majority of twentieth century mathematicians accept Cantor's theory. There are, however, some exceptions.

Brouwer (1882–1966)

L. E. J. Brouwer was one of the founders of the school of 'intuitionism'. For intuitionists, numbers do not exist independently of human knowledge.

---

[29] If A is the set of natural numbers, the union of the power sets in the above sequence is a set which is more infinite than any of those power sets. It is called the 'least Zermelo universe'.

[30] See G. Kreisel, 'Informal Rigour', in *The Philosophy of Mathematics,* ed. J. Hintikka (London: Oxford University Press, 1969), pp. 87-88.

When an intuitionist makes some assertion about numbers, he or she understands it to imply that the truth of that assertion is within the compass of present human knowledge. More precisely, a proposition p, about numbers, is true only if we know of some procedure whereby finite humans can determine that p is true. Thus, for the intuitionist, not-not-p does not entail p, since not-not-p is understood to be, not so much a denial of not-p itself, as a denial that the truth of not-p is within the range of current human knowledge.

We have no way of knowing that not-p

does not entail

We have some way of knowing that p

and so, for the intuitionist, not-not-p does not entail p.

For an intuitionist, a set can be non-infinite, in the sense that there is currently no way for limited human beings to know that it is infinite, without it being the case that the set is finite, in the sense that there is currently some way for limited human beings to know that it is finite. For example, it has been conjectured, but not yet proved, that the set of twin primes is infinite.[31] For most mathematicians, this conjecture is either true or false. The intuitionist rejects this dichotomy, claiming that, given the present human ignorance of the matter, the conjecture is not true, and its negation is not true either. The set of twin primes is not infinite, nor is it finite.

H. Eves describes intuitionism as follows:

> The intuitionist thesis is that mathematics is to be built solely
> by finite constructive methods on the intuitively given sequence
> of natural numbers.[32]

The methods are 'constructive' because we humans create the mathematics, and they are 'finite' because we creators of mathematics are assumed to be finite.

Eves considers an example due to Brouwer. If the sequence

$$123456789$$

occurs in the decimal expansion of π, let k be the number of the decimal place where its first occurrence begins, and let $x = (-1)^k$. Otherwise,

---

[31] Two primes are 'twins' if they differ by 2. For example, 11 and 13 are twin primes.
[32] H. Eves, *An Introduction to the History of Mathematics*, 5th edn. (Philadelphia: Saunders, 1983), p. 480.

let $x = 0$. We do not know, at the present time, whether the sequence 123456789 occurs in the decimal expansion of π, and, as far as we know at the moment, we have no finite constructive method that would help us settle the matter. Since the intuitionist believes that the decimal expansion of π does not exist independently of human calculation, he or she believes that the number x is not well-defined.[33] For the intuitionist, the proposition

$$x = 0$$

is not true, and the proposition

$$x \neq 0$$

is not true either. Eves remarks:

> for the intuitionist, the law of the excluded middle holds for finite sets but should not be employed when dealing with infinite sets

— such as the sequence of digits in the decimal expansion of π.[34] The reason the law of the excluded middle holds for finite sets is that finite sets are accessible to finite human knowledge. The reason that law should not be employed with infinite sets is that, in general, they are not accessible to finite human knowledge.

Eves reports that when Brouwer was an editor of the prestigious *Mathematische Annalen*, he refused

> all papers that applied the law of the excluded middle to propositions whose truth or falsity could not be decided in a finite number of steps.[35]

Robinson (1918–1974)

With his nonstandard analysis, Abraham Robinson gave the first rigorous treatment of the infinitesimal. One might think that this would have inclined him to be a friend of the infinite, but, although he was not an intuitionist, he rejected it outright:

---

[33] Actually, as reported in *The Times* of January 22, 1996, we now do know that 123456789 occurs, several times, in the decimal expansion of π. Thus Brouwer's example is obsolete. However, exactly the same point can be made by, say, taking the sequence 123456789123456789.

[34] Eves, p. 481.

[35] Ibid.

> Infinite totalities do not exist in any sense of the word (i.e.,
> either really or ideally). More precisely, any mention, or pur-
> ported mention, of infinite totalities is, literally, *meaningless*.[36]

This position allowed Robinson to cut the Gordian Knot of the Continuum
Hypothesis: rather than being 'upset' about whether it was true or false,
he simply dismissed it as 'meaningless'.[37]

Moore (1990)

A. W. Moore accepts the set of natural numbers, but he belittles its infinitude, calling it finite. Thus Moore claims that 'whatever exists is finite' and
'nothing at all is mathematically infinite'.[38] Furthermore, taking himself
as a typical human being, Moore asserts he is finite in the following senses:
(1) he is 'cast into' a world not of his own making;
(2) a minor rearrangement of the particles inside his body would be enough
to annihilate him;
(3) he is a finite, 'parcel of matter';
(4) he cannot receive the 'metaphysically infinite whole' without being de-
stroyed in the process;
(5) what he does receive he must receive at particular times;
(6) whatever he receives points to other things he has not received.

Moore does not give any arguments for these gloomy claims. Indeed,
in relation to (2), he admits that this 'begs the question of whether I have
an afterlife'. Since Moore's whole list is question-begging, one may as well
consider an optimistic point of view, consisting of some or all the following
theses:
(1) humans can shape their world, making it conform to their free decisions;
(2) humans are immortal;
(3) humans are immaterial souls, capable of learning an infinite number of
truths;
(4) humans can receive God, becoming close to him without being absorbed
by him;
(5) humans can transcend time in the contemplation of eternal truths;
(6) humans can survey infinite totalities in mathematics.

Moore believes his claims are not gloomy.[39] For example, he does not

---

[36] A. Robinson, 'Formalism 64', in *Logic, Methodology and Philosophy of Science*, ed.
Y. Bar-Hillel (Amsterdam: North-Holland, 1965), p. 230.
[37] Ibid., p. 232.
[38] Moore, *The Infinite*, pp. 222 and 224.
[39] Private letter to W. S. Anglin, dated 13 January, 1996.

think that not living forever is gloomy. In support of Moore, a Buddhist might point out that, if we live forever, we shall never get to the non-existence of nirvana. Against Moore, a Christian might point out that the exaltation of non-existence is a substitution of 'darkness for light'. It was Paul who was right when he said that, if only in this life we have hope, 'we are to be pitied' (1 *Corinthians* 15:19).

Scepticism

How sceptical should one be of the infinite? In one way it is always safer to be sceptical: you are not prone to error, and you will not be disappointed on account of having held overly optimistic beliefs. On the other hand, the sceptic is in danger of becoming dogmatically entrenched in his or her scepticism, and forfeiting many beautiful, useful, and ennobling truths.

Chapter 3

Mathematics and Infinity III

In Chapter 1 we gave four arguments of Aristotle against the actually infinite. In this chapter we give seven arguments in its favour.

3.1 Considerations from Contemporary Mathematics

In the twentieth century we surpassed the mathematical achievements of every previous century. We did this, moreover, using a mathematics that is based on, and permeated by, Cantor's infinite sets. It was only by means of employing the infinite that we succeeded in erecting the magnificent structures now found in logic, algebra, analysis, topology, and other branches of mathematics. It was only by means of employing the infinite that we were able to solve the Bachet equation — just to mention one of many problems that had long resisted solution but were finally disposed of, using the infinite, in the twentieth century.[1] If success and fruitfulness are ever, in any way, evidence for the validity of an idea, then they have surely justified the infinite in mathematics. The simple truth is this: in contemporary mathematics we do think in terms of the infinite, and, as a result, we

---

[1] If k is a given nonzero integer, and x and y are variables ranging over the integers, the *Bachet* or *Mordell* equation is the equation $y^2 = x^3 - k$. In other words, the problem is that of finding cubes and squares that differ by a given integer, and then proving that one has found them all. It was first solved by Alan Baker in 1968. His solution makes use of complex analysis, a branch of mathematics that relies on infinite sets and infinitistic methods.

engage in mathematical activity of hitherto unparalleled breadth, depth, and practical application. The few mathematicians who abhor the infinite are outsiders to this accomplishment. When intuitionist Arend Heyting says that the wonderful mathematical achievements based on the infinite are 'noxious ornaments' that should be excised from mathematics, this is simply sour grapes.[2]

3.2 Considerations from the History of Mathematics

History shows that sometimes mathematicians make important advances only because they are not sceptical of the infinite. Archimedes gave the first complete treatment of the formula for the area of a circle only because he went beyond Euclid's finitistic restrictions on 'constructible' segments, and simply assumed the existence of a segment with length π. Newton and Leibniz produced the calculus only because they were willing to accept the nonfinitistic nature of infinitesimals and integrals. Cantor gave us a foundation for topology and analysis only because he accepted the existence of nonconstructible infinite sets. By taking an infinitist view of mathematics, mathematicians have often achieved results which they could not have obtained if they had accepted the limitations recommended by Aristotelians, empiricists, or intuitionists.

3.3 Considerations from Physics

Contemporary physics gives us a good understanding of many phenomena in the physical world. In particular, it gives us a good understanding of motion and electricity.

As for motion, this understanding presupposes the existence of infinitely many positions and instants in space-time. As for electricity, this understanding is in terms of advanced calculus and complex analysis — and hence in terms of the infinite which goes with both these branches of mathematics.

If, therefore, the belief that we can comprehend the infinite is false, one of the pillars supporting our understanding in physics is cut away, and much of our so-called insight in physics is delusion. Since this is not the case, we may conclude that we can understand things via the infinite, and we are thus, in a sense, infinite ourselves.[3]

---

[2] A. Heyting, 'Disputation', in *Philosophy of Mathematics,* 2nd edn., ed. P. Benacerraf and H. Putnam (New York: Cambridge University Press, 1983), p. 74.

[3] Note that this argument does not presuppose the view that there is some 'true'

3.4 Considerations from Logic

The following argument was discovered by John Lucas, and is now defended by Roger Penrose. It has elicited passionate emotion from certain academics who have, more than once, angrily declared it dead. Penrose has published a book in its defense — namely, his *Shadows of the Mind* — and, at a conference held in Oxford on May 25, 1996, he and Lucas listed the various, often repeated, objections to the argument, and rebutted them, one by one.

The argument is based on two premisses. The first is that the human mind is basically consistent, and the second is that, given any consistent, logical system, strong enough to contain arithmetic, there is a sentence in it which we can know to be true, but which the system itself cannot prove. The conclusion is that the human mind transcends any logical system. In particular, it is not a mere computer. And it is a corollary to this that the human mind is not subject to any finite set of rules.

In defense of the first premiss, note that it would be self-defeating for any philosopher to deny it. If our minds are inconsistent, we are not rational creatures, and we can only fail if we try to explore reality by means of rational inquiry.

In defense of the second premiss, there is the well-known Gödel Incompleteness Theorem. This is too complicated to present here in all its details, but a rough sketch of it — and how it is used in our argument — can be given as follows.[4]

Suppose we have a consistent, formal, axiomatic system S, strong enough to include arithmetic. Suppose we number the well-formed propositional formulas of S (by, say, listing them in some alphabetical order). Somewhere on the list we shall find $x + 2 = 3$ and somewhere else we shall find $x + y \neq y + x$. Now, as Gödel showed, one of the well-formed propositional formulas on the list is equivalent to

> there is no proof in S whose conclusion is that the counting number x satisfies formula x.

(A number a 'satisfies' a formula Px just in case Pa is true. For example, 1 satisfies the formula '$x + 2 = 3$'.) Suppose Gödel's formula is number g on the list of well-formed propositional formulas. Now either g satisfies formula g or it does not.

---

theory of the physical universe which involves the infinite. All the argument is saying is that we are infinite in the sense that we have a good (not a perfect!) understanding of some aspects of the physical universe — and this in terms of the infinite.

[4] For a more detailed account, see the chapter in Anglin and Lambek, *The Heritage of Thales* (New York: Springer, 1995).

If g does not satisfy formula g, then there *is* a proof in S whose conclusion is that g satisfies formula g. Hence, assuming S is consistent, g satisfies formula g. Contradiction. Hence g does satisfy formula g. But then there is no proof that it does. We can see that the statement

> g satisfies formula g

is true, but there is no corresponding proof of this in system S. So much for the second premiss.

Now, to obtain a contradiction, suppose our minds can be identified with some particular, formal system S which includes arithmetic — the way a computer can be identified with a formal system expressing the precise rules governing its outputs. Since our minds are consistent, this system is consistent too. Let p be a statement of S which is not provable in S but which we none the less realise is true. Since p is not provable, the system S itself cannot vouch for the truth of p. However, we can. Contradiction.

To put it another way, given any formal system that might represent the human mind, there is something about that system which we, at least in principle, can correctly conclude to be true, but which necessarily goes beyond the proof capacities of that system.

Hence our minds cannot be identified with any given formal system. Our minds work in a way that is above and beyond formal systems, and, in this sense, we are infinite.[5]

3.5 Considerations from Psychology

The following argument was put forward by Dedekind and endorsed by Royce. It begins with the premiss that human psychology is *reflexive* — in the sense that, no matter what we are thinking, we can think about ourselves thinking that thing. For example, Douglas can think about his thinking too much about Marilyn, and he can think about his thinking too much about thinking too much about Marilyn, and so on — without any limit on the number of possible levels of introspection.[6]

Let T stand for

> Douglas thinks too much about the fact that ...

and let p stand for

> Marilyn has lovely brown eyes.

---

[5] See R. Penrose, *The Emperor's New Mind* (New York: Oxford University Press, 1989).

[6] See J. Royce, *The World and the Individual* (New York: Macmillan, 1912), p. 511.

Then TTp stands for

> Douglas thinks too much about the fact that Douglas thinks
> too much about the fact that Marilyn has lovely brown eyes.

In order to express the depths of Douglas's introspection, we define $T1p$ as Tp, and $T(n+1)p$ as $T(Tnp)$ — where n is a positive integer. For example, $T2p$ is TTp. Having visited a hundred different psychoanalysts, and discussed with each one what he had discussed with the previous ones, Douglas has become very introspective indeed:

> Well, Dr. Z, I told Dr. Y about the fact that $T99p$.
> — I see. Let's work on that fact for a few sessions. How do you
> feel about it?
> — I'm afraid that it will soon be the case that $T100p$.

Douglas and his latest psychoanalyst then discuss the possibility that the analysis might go on forever.

To this it may be objected that there is a limit on Douglas's memory or on his life span which ensures that it could never be the case that, say, $T1000000p$.

However, is there? Although it is true that there are only finitely many atoms in Douglas's brain, this does not imply that there is an upper limit on the complexity of his memories. For the atom is divisible, and it may be infinitely divisible, and thus capable of supporting configurations of any desired degree of complexity. Moreover, Douglas may be a soul rather than a brain, and it may be that souls can contain unlimited quantities of information. Even Douglas's life span need not be an obstacle. For Douglas may learn how to move from Tnp to $T(n+1)p$ in just $1/2^n$ years. If so, he will be able to get through the whole sequence $T1p$, $T2p$, $T3p$, ...in just

$$1 + \frac{1}{2} + \frac{1}{4} + \frac{1}{8} + \cdots$$

$= 2$ years. Or Douglas may be immortal.

In any case, granted the premiss that we can always think about our thoughts, human beings are infinite in the sense that there is no upper limit on the quantity or complexity of those thoughts.[7]

---

[7] Someone who wants to reject the conclusion of an argument can always reject the premiss, complaining that no one would accept that premiss unless he or she already agreed with the conclusion. This complaint — sometimes accompanied by cries of 'petitio principii' — is not legitimate unless the premiss is virtually identical with the conclusion.

3.6 Considerations from Epistemology

The 'scientific method' involves an interplay of observation, hypothesis forming and deduction, which enables us to come to know the physical world. On the basis of experiments, we conjecture the existence of certain laws, and we subject our conjectures to further experiments which either confirm or refute them. In this way we acquire knowledge about the natural world. Admittedly, what we learn this way is subject to refinement, but, in an ordinary sense of the word 'know', by using the scientific method, we come to know the laws of the material universe.

Indeed, if we were dropped in a wholly different universe, with physical components and natural laws different from those of the actual universe, the scientific method would still work. It is not possible that a universe (e.g. our own) be so recondite that it would not yield its secrets to the scientific method. By means of observation, hypothesis forming, and deduction, the human mind could discover the laws of any universe, no matter how complicated. This is a basic presupposition of science.

Furthermore, it follows from this presupposition that the human mind is infinite in the sense that it can discover laws of nature regardless of their complexity. There is no upper limit on our ability to penetrate the subtleties of the material world.

Against this conclusion, a sceptic might argue that what we come to 'know' using the scientific method is at best an approximation to the truth, and an approximation within the compass of a finite mind. This is a good point, but it is an unduly pessimistic point. The fact that gold has a significantly greater density than water, and hence sinks in water, may be 'a mere approximation to the truth', but it is also a very good approximation to the truth. And if the human mind, by following the scientific method, can find equally good approximations, regardless of the level of complexity of the situation under study, then it is, in a sense, infinite.

3.7 Considerations from Theology

The orthodox theist is committed to God's omniscience. God knows every true proposition.[8] In particular, he knows true mathematical propositions such as

$$2 + 5 = 7$$

and

---

[8] W. S. Anglin, *Free Will and the Christian Faith* (Oxford: Clarendon Press, 1990), ch. 2 and 4.

If Euclid's axioms are true then the medians of a triangle are concurrent.

Now consider Goldbach's Conjecture that every even number greater than 2 is the sum of two primes. (For example, $10 = 3 + 7$.) For the sake of argument, let us suppose that there is neither a counterexample to Goldbach's Conjecture nor a proof of Goldbach's Conjecture (from the usual axioms of arithmetic).[9] On the view of mathematics held by most mathematicians, there is an infinite set S of sums of two primes:

$$S = \{x + y \mid x,\ y \text{ are prime}\}$$

Furthermore, assuming this is so, God can examine this set S and determine that there is no even number greater than 2 missing from it.[10] On a strict finitist view, however, there is no infinite set for God to examine, and since, moreover, there is no proof (we are assuming) of Goldbach's Conjecture, it follows that God cannot know whether every even number greater than 2 is a sum of two primes.[11]

We thus have two pictures of God. In the first picture there are infinite totalities in mathematics, and God has a knowledge of them. In the second picture there are no infinite totalities in mathematics, and God cannot be held to know whether, say, Goldbach's Conjecture is true — because there is nothing true to know.[12]

A theist, at any rate, must reject the second picture. For it contradicts the glory and the majesty of God. It belittles him. In the second picture, God's ignorance is excused, but he is none the less ignorant. It is only the first picture that accords with the fact that God is great and worthy of our praise. For a theist, God can create infinitely many stars, he can know infinitely many facts, and he can love us for infinitely many years. Those who deny every sort of infinite implicitly denigrate God's power.[13]

---

[9] The situation we are hypothesising is a reasonable one because of Gödel's result that some statements in arithmetic are undecidable (assuming arithmetic is consistent).

[10] If there were an even number missing from it, Goldbach's Conjecture would have a counterexample, which, we are assuming, it does not.

[11] This does not necessarily conflict with God's omniscience, since, on a strict finitist view, there is nothing here to know.

[12] M. Dummett supports the strict finitist picture in *The Logical Basis of Metaphysics* (Cambridge, Mass.: Harvard University Press, 1991), pp. 348-51.

[13] See 1 *Corinthians* 13:8; *Ephesians* 3:20; *Hebrews* 1:12. Note that the statement 'God can create infinitely many stars' does not imply the statement 'if there is a God, then there are infinitely many stars'.

3.8 Summary

From the above, we see that a person who rejects the infinite
(1) cannot participate in the triumph of contemporary mathematics,
(2) cannot be in tune with the role of the actually infinite in mathematical innovation,
(3) cannot recognise the success of contemporary physics,
(4) cannot comprehend a natural corollary of Gödel's Incompleteness Theorem,
(5) cannot accept the reflexivity principle of psychology,
(6) cannot endorse the full power of the scientific method, and
(7) cannot give due glory to God.

3.9 Appendix
Considerations from Irrationality

The following argument is not a very strong one, but it does have some pleasant mathematical features.

> If a line segment is not made up of infinitely many points, but only a finite number, then each of those points will have some finite length $\epsilon > 0$.
> So every line segment will have a length of the form ϵn, where n is a natural number.
> Thus, without loss of generality, we can suppose that the legs of an isosceles right triangle each measure ϵa, where a is a natural number.
> By the Theorem of Pythagoras, the hypotenuse measures $\epsilon a \times \sqrt{2}$.
> Also, given our assumption, this hypotenuse measures ϵb, for some natural number b.
> Hence $\epsilon a \times \sqrt{2} = \epsilon b$.
> Thus $\sqrt{2} = b/a$ — a rational number.
> Contradiction.
> Hence a line segment is made up of infinitely many points, and there are infinites in mathematics.

> Furthermore, it will not help the finitist to say that the points in a line segment come in two different sizes ϵ_1 and ϵ_2.
> For suppose this is the case, and suppose that the line segment

of length 1 contains a_1 points of size ϵ_1 and a_2 points of size ϵ_2 — where a_1 and a_2 are natural numbers.

Using the Theorem of Pythagoras on a right triangle with legs 1 and 1, we get

$$\sqrt{2}(a_1\epsilon_1 + a_2\epsilon_2) = b_1\epsilon_1 + b_2\epsilon_2$$

and hence

$$\frac{\epsilon_1}{\epsilon_2} = \frac{b_2 - \sqrt{2}a_2}{\sqrt{2}a_1 - b_1}$$

Using the Theorem of Pythagoras on a right triangle with legs 1 and 2, we get

$$\sqrt{5}(a_1\epsilon_1 + a_2\epsilon_2) = c_1\epsilon_1 + c_2\epsilon_2$$

and hence

$$\frac{\epsilon_1}{\epsilon_2} = \frac{c_2 - \sqrt{5}a_2}{\sqrt{5}a_1 - c_1}$$

Thus

$$-b_2c_1 + a_2c_1\sqrt{2} + a_1b_2\sqrt{5} = -b_1c_2 + a_1c_2\sqrt{2} + a_2b_1\sqrt{5}$$

Since $\sqrt{5}$ is not a member of $\mathbf{Q}(\sqrt{2})$, it follows that $a_1b_2 = a_2b_1$. Hence from the fact that

$$\sqrt{2}(a_1\epsilon_1 + a_2\epsilon_2) = b_1\epsilon_1 + b_2\epsilon_2$$

it follows (from multiplying both sides of the equation by b_2) that

$$\sqrt{2}a_2 = b_2$$

again against the irrationality of $\sqrt{2}$.

Chapter 4

Mathematics and Metaphysics

The goal of this chapter is to discuss some answers to the question

> do mathematical objects exist, and, if so, what kind of existence
> do they have?

Answers to related but distinct questions are not dealt with in this chapter. For example, the term 'formalism' is sometimes used as a label for the view that the objects of mathematics are certain chalk or ink marks. This is an answer to the above question, and it is covered in this chapter. The term 'formalism' is also used as a label for the view that the objects of mathematics are meaningless deductive games — without this implying anything about the kind of existence these games have (if any). This kind of 'formalism' does not give an answer to the above question, and it is not in the scope of this chapter.

4.1 Mathematical Objects

To begin, let us say something about the meaning of the term 'mathematical object'. For some people, mathematical objects are sets or classes. For others, they are positions in 'structures', or even the structures themselves. For some, mathematical objects are definite, and hence, for example, each set is either finite or infinite. For others, however, they are vague, and some sets are neither finite nor infinite. For some people, all mathematical objects are abstract. However, for others, all objects, including mathematical objects, are concrete, physical bodies.

In order not to prejudice any issues, we use the term 'mathematical object' in as broad a way as possible. In particular, we use it in a way that is broad enough to include the following as 'mathematical objects': (a) the number 2, (b) the natural number system, (c) the triangle with sides 3, 4, and 5, and (d) the Euclidean plane.

Proponents of the view that all mathematical objects are sets or classes sometimes ask *which* mathematical object 2 is. Is it the class of all pairs, or is it the set $\{\phi, \{\phi\}\}$, or is it the set $\{\{\phi\}\}$? Since choosing one of these alternatives does not make any difference to number theory, some people have concluded that there is no truth of the matter (2 is vague), while others have concluded that, although there is a correct answer to the question, it is unknowable and unimportant. A structuralist might say that 2 is the class of all ordered pairs $(M, 2_M)$, such that M is a model of second order Peano arithmetic and 2_M is the successor of the successor of 0 in that model.

One can also ask tricky questions about the 3-4-5 triangle. For example, 'does the bisector of its right angle pass through the origin?' Someone who thinks that some mathematical objects are vague might reply that it neither does nor doesn't. Someone else might answer that 'the 3-4-5 triangle' is actually, not a triangle, but a set — namely, the set of all 3-4-5 triangles — and hence it does not have any particular orientation. Or perhaps 'the 3-4-5 triangle' is a variable x ranging over the set of all 3-4-5 triangles. Given classical logic, we could think of this x as having unknown but definite properties: either the bisector of x's right angle passes through the origin or it does not.

As we shall see in what follows, the view that mathematical objects are vague fits with ontological positions such as 'intuitionism', while the view that they are definite fits with ontological positions such as 'formalism' and 'Platonism'.

4.2 The Fundamental Choice

A metaphysician who wants to have definite opinions on the ontological status of sets, numbers, modular groups and the like will have to begin by making a fundamental choice between the following options:

(A)   There is no sense in which any mathematical object exists.

(B)   (A) is false.

Position (A) is *mathematical nihilism* or *fictionalism*. H. Field is a nihilist in this sense.[1]

---

[1]H. Field, *Science Without Numbers* (Princeton NJ: Princeton University Press,

4.3 Nihilism and Truth

Suppose Grace is thinking about a square. There are various ontological possibilities. First, she might be thinking about something which exists independently of the human mind. Second, she might be thinking about something which exists only because people think about it: that is, it might be a mental artefact. Third, she might be thinking about something which does not exist at all, something which lacks even a mental existence. If, for example, she is thinking about a round square, then there is a sense in which such an object does not even 'exist in the mind'. What we mean by mathematical nihilism is this rather strong third possibility: squares, of whatever kind, have no existence of any sort.

Nihilism provokes a number of questions.

1. How is it that we can use nonexistent objects, such as the natural numbers, to explain and control the physical world?

2. How is it that mathematical words refer to objects that are not there to be referred to?

3. How is it that a statement, such as $2 + 2 = 4$, can be a paradigm of truth when, according to the nihilist, it is a mere fiction?

Since, for the nihilist, mathematical statements refer to nonexistent entities, the nihilist might choose to regard all mathematical statements as false. A theorem such as

> There are infinitely many primes

is actually false if there are no numbers at all. Hans Vaihinger is an example of a nihilist who thinks that, indeed, all mathematical statements are false.[2]

If the nihilist wants to hold a less extreme view about truth in mathematics, perhaps the best position from which to do so is 'postulationism' — sometimes called 'if-thenism'. In other words, someone who has chosen option (A) might be well advised to pick the more specific option

(C)   *Postulationism:* nihilism is true, but we can still play the game of pretending that a mathematical object exists if a set of axioms implies that it exists.

> For example, the statement

> There are infinitely many primes

1980). I spoke to Professor Field at the Truth in Mathematics conference held at Mussomeli in 1995, and he agreed with me that he was a nihilist.

[2] Hans Vaihinger, *Philosophy of the As-If.* (London: Routledge and Kegan Paul, 1935).

should be counted as true just in case there is a proof, based on the relevant axioms, whose conclusion is 'There are infinitely many primes'. For the postulationist, truth in mathematics has to do, not with numbers or sets, but with entailments between statements.

In some cases there is a statement which is not determined by the axioms. It holds under one interpretation of the axioms, but its negation holds another another interpretation. For the postulationist, such a statement is not true and its negation is not true either. For example, relative to the other axioms of Euclidean geometry, the Parallel Postulate — asserting the existence of a point where two straight lines meet if they are tilted towards each other — is not determined. It holds under the standard interpretation of 'straight line', but, under a 'hyperbolic interpretation', it does not. For the postulationist, given only 'Euclid's other axioms', the Parallel Postulate is not true and its negation is not true either.

Postulationism is sometimes associated with a view called *structuralism* which sees mathematics as an interplay and comparison of 'patterns'. For a structuralist, what is important is not the metaphysics of, say, UFD's and PID's, but the fact that every structure of the first kind is *a fortiori* a structure of the second kind. Postulationism is not, however, the same as structuralism, since structuralism is compatible with realist views on mathematical objects. The fact that one puts the emphasis on structures does not have any specific implications for one's beliefs about the ontological status of the structures.[3]

4.4 A Second Choice

If the metaphysician eschews nihilism, the next choice is between the following views:

(D)   *Mathematical idealism:* there are mathematical objects, but they exist only in the human mind.

(E)   *Mathematical realism:* there are mathematical objects, and at least some of them exist independently of the human mind.

Note that realism includes many possibilities. It includes the possibility that most, but not all, mathematical objects are found only in human minds. It includes the possibility that all mathematical objects exist in, and only in, minds, but these minds are not all human. It includes the possibility that all mathematical objects exist, not in human minds, but in human

---

[3] We say more about structuralism at the end of this chapter.

brains. And, of course, it includes the possibility that all mathematical objects exist independently of any mind whatsoever. We shall sort out the contents of this olio, but, first, let us take a closer look at mathematical idealism.

4.5 Idealism

Mathematical idealism exalts the human factor in mathematics, and, in so doing, explains its artistic side, its cultural aspect, and its immediacy in the human situation.

One of the weaknesses of mathematical idealism is that it fails to deal with the feeling that counting numbers form part of the logical framework of possibility, even in possible universes where humanity is not instantiated. How is the mathematical idealist going to explain the truth of a counterfactual such as

> Even if there were no human beings, the number of planets circling the sun would still be a square number ?

In any case, if the metaphysician chooses idealism, then one of his or her next options is between the following:

(F)   *Finitist idealism:* mathematical idealism is true, and, moreover, the human mind is finite in such a way that no mathematical object exists unless it can be 'constructed' by the human mind by means of a finite sequence of descriptions, or calculations, or other finite mathematical manipulations.

(G)   *Infinitist idealism:* mathematical idealism is true but (F) is false.

P. Ernest's 'social constructivism' is a version of infinitist idealism. Ernest holds that mathematical objects (including infinite sets) are constructed by mathematicians, and exist in their collective consciousness. Hence

> if all humans and their products ceased to exist, then so too would the concepts of truth, money and the objects of mathematics. Social constructivism therefore involves the rejection of platonism.[4]

---

[4] P. Ernest, *The Philosophy of Mathematics Education* (London: The Falmer Press, 1991), p. 57.

For Ernest, if there are two separate groups of mathematicians, with one group accepting the existence of, say, the set of reals, and the other rejecting it, then both groups have 'objective mathematical knowledge' about the set of reals.[5] There is no need to analyse the metaphysical presuppositions of these groups, or to ask after the requirements of reason: it is enough that there are two groups of mathematicians, and then they are both right, even if they contradict each other.

The most popular form of idealism in the philosophy of mathematics is, not social constructivism, but a form of finitist idealism called 'intuitionism'.

4.6 Intuitionism

Intuitionism is a view according to which mathematical objects are concepts in finite, human minds. The mathematician intuits the basic mathematical objects, such as the natural numbers, and then, using some finite procedure, mentally constructs other mathematical objects from these basic objects. Note that the intuition of basic mathematical objects is *not*, for the intuitionist, some kind of perception of something outside the human mind.

According to Brouwer, one of the founders of intuitionism:

> The ... point of view that there are no non-experienced truths and that logic is not an absolutely reliable instrument to discover truths has found acceptance with regard to mathematics much later than with regards to practical life and science. Mathematics rigorously treated from this point of view, including deducing theorems exclusively by means of introspective construction, is called intuitionistic mathematics.[6]

Another intuitionist, A. Heyting, asserts:

> mathematical objects are by their very nature dependent on human thought. Their existence is guaranteed only insofar as they can be determined by thought. They have properties only insofar as these can be discerned in them by thought. But this possibility of knowledge is revealed to us only by the act of knowing

---

[5] Ibid., p. 63.

[6] L. E. J. Brouwer, 'Consciousness, Philosophy, and Mathematics', in *Philosophy of Mathematics,* p. 90.

itself. Faith in transcendental existence, unsupported by concepts, must be rejected as a means of mathematical truth.[7]

As Wittgenstein suggests, there is an analogy between the mathematician, at least as seen by the intuitionist, and the storyteller.[8] The mathematician intuits natural numbers; the storyteller recollects cultural symbols. Logical norms guide the mathematician; cultural patterns direct the storyteller. The mathematician produces theories which help us understand the physical world; the storyteller relates tales which help us comprehend the human soul.

Both storyteller and mathematician leave certain possibilities open. Since the storyteller cannot construct an absolutely complete account of Hercules, there are statements p about Hercules such that neither p nor its negation is true. An example might be the statement

Hercules has a mole on the fifth finger of his right hand.

Similarly, since the mathematician cannot construct a complete decimal expansion for π, there are some statements p about π such that neither p nor its negation is true. An example might be

The decimal expansion of π contains the sequence 123456789999.

The intuitionist's view thus leads to the denial of the Law of the Excluded Middle: p or not-p.[9]

For the storyteller, the universe in which Hercules does his deeds is expandable to any size required by the tale. There is no particular upper bound on its dimensions. Yet it would be going too far to say that its volume is infinite. The statement

The universe in which Hercules does his deeds is infinite

is best categorised, not as true, but as 'going beyond the story'. Analogously, in mathematics, although there is no particular upper bound on the set of reals, it would be going too far to assert that it is actually infinite. For the intuitionist, there is no mathematical object which cannot be constructed in a finite number of steps.[10]

---

[7] A. Heyting, 'The Intuitionist Foundations of Mathematics', in *Philosophy of Mathematics,* p. 53.

[8] L. Wittgenstein, *Remarks on the Foundations of Mathematics* (Cambridge: The MIT Press, 1956), p. 138e.

[9] When someone suggested that the law of the excluded middle is true because God, at least, knows the actual state of affairs, Brouwer replied, 'I do not have a pipeline to God'. Intuitionists reject prophecy. See p. 184 in Constance Reid, *Hilbert* (New York: Springer-Verlag, 1970).

[10] S. Lavine, *Understanding the Infinite,* pp. 171-174. Brouwer sometimes *said* he accepted the set of natural numbers, but such remarks were in conflict with his general finitism.

Intuitionism has some impressive strengths. For example, it does not get bogged down in metaphysical speculations about the objective truth of undecidable statements concerning the infinite.

However, intuitionism also has weaknesses. Some are due to its being a form of idealism, and some to its being a form of finitism. Holding that human intuition is basically finite, and believing that an entity does not exist unless it can be experienced by humans, intuitionists reject mathematical objects which cannot be constructed in a finite manner. The upshot of this is that intuitionists cannot prove many standard mathematical results. For example, they lack the equipment needed to establish some of the basic theorems of calculus. It is because it excludes many important branches of mathematics (not to mention physics) that most mathematicians reject intuitionism.

Another weakness of intuitionism is that it does not do justice to the fact that mathematicians often experience the achievement of a new result as a discovery. No one says that the Mesopotamians or Chinese *invented* the Theorem of Pythagoras. What we all say — those of us who know our history — is that the Mesopotamians or Chinese *discovered* the Theorem of Pythagoras. Again, no one says that the great dodecahedron was *composed* in 1809. What we all say is that it was *discovered* in 1809 (by Louis Poinsot). According to any reasonable Logic of Discovery, the phrase

> Person A discovered X

entails the phrase

> X was there waiting to be discovered

— and this conclusion is one which practising mathematicians often endorse.

4.7 A Third Choice

Let us return to realism. We could also begin here with a distinction between finitists and infinitists, but a more natural starting point is the distinction between the following positions:

(H)   *Mathematical materialism:* There are mathematical objects which exist independently of human minds, and all these objects are material (i.e. physical) objects.

(I)   *Mathematical immaterialism:* mathematical realism is true, and at least some of the mathematical objects existing independently of the human mind are not material objects.

4.8 Mathematical Materialism

Mathematical materialists enjoy all the advantages of realism: for example, they have no particular difficulty with 'discovery', with nonconstructive methods, or with the Law of the Excluded Middle. Their ontology, moreover, is attractively simple: numbers are in the same category as trees.

One *difficulty* for mathematical materialism is the feeling that, unlike material objects, numbers are not contingent. For example, it is not only true, but necessarily true, that

> There is a prime between 10 and 20.

Even in possible universes devoid of matter this is still the case.

A second difficulty for mathematical materialism is the truism that, although the chalk mark on the blackboard has a nonzero mass, a nonzero thickness, a nonzero potential energy, a finite length, and a finite duration, the mathematical line it represents has no mass, no thickness, no potential energy, an infinite length, and a timeless duration. These latter qualities are not qualities of a physical object. Nor is this truism restricted to lines and chalk marks: it applies to any mathematical object and its physical representations.

A third difficulty for materialism is that the quantity of matter is limited, whereas the number of mathematical objects is not. Let A be the (possibly transfinite) cardinality of the set of physical objects. Then, for the materialist, no set of mathematical objects has a cardinality larger than A (since any set of mathematical objects is a subset of the set of physical objects). Yet, given standard assumptions from set theory, there is no upper bound (finite or transfinite) on the cardinalities sets can have.

Two important sub-species of mathematical materialism are the following:

(J)   *Formalism:* The objects of mathematics are the material signs used to represent them in mathematical literature.

(K)   *Concretism:* The basic objects of mathematics are concrete, non-linguistic objects, such as apples.

Concretism has had some supporters. In his *System of Logic,* J. S. Mill (1806–1873) argues that 'there are no such things as numbers in the abstract'.[11] On the contrary,

---

[11] J. S. Mill, *System of Logic* (London: Longmans, 1961), p. 167.

Each of the numbers two, three, four, &c., denotes physical phenomena.[12]

Geometry, too, 'is a strictly physical science'.[13]

Since concretism is an unusual view in the ontology of mathematics, we shall not say more about it here. The reader is invited to consult P. Maddy's defense of a version of it in her book *Realism in Mathematics.*

4.9 Formalism

Formalism is the view that the objects of mathematics are physical strings of characters (or signs). In the words of D. Hilbert (1862–1943),

> the subject matter of mathematics is ... the concrete symbols themselves.[14]

For example, as H. Lebesgue asserts, the integers are 'material symbols intended to give reports of physical counting experiences'.[15]

On this view, doing mathematics is manipulating physical strings of characters, according to certain rules. For the formalist, the number 1 should be understood in terms of the numerals for 1, and the 'grammar' which dictates their use. Mathematical objects are thus reduced to ink marks and sound wave patterns, and, ultimately, to quarks and force fields.

Formalism has some difficulties all of its own. First, it is difficult to explain what is meant by 'mathematical symbols', especially if one wants to avoid defining them in terms of mental concepts or abstract universals. Is the number 1 a collection of ink marks which grows larger with time? Is it the set of patterns which typographers currently use as numerals for 1? Does it include the 1-shaped birthday cake Professor Smith baked for his niece Melinda? The formalist needs a kind of 'type-token' distinction, but, precisely, this distinction requires reference to non-material objects.

A second weakness of formalism is that it sees mathematics as a language without meaning. The 'words' need not have reference; the 'discourse' need not have purpose or utility. To quote Hilbert, 'these numerical symbols which are themselves our subject matter have no significance in

---

[12] Ibid., p. 400.

[13] Ibid., p. 403.

[14] D. Hilbert, 'On the Infinite', in *Philosophy of Mathematics,* 2nd edn. p. 192. In other places Hilbert said other things, and it would be wrong to say he was a consistent materialist. See, for example, page 186 of Constance Reid, *Hilbert,* (New York: Springer-Verlag, 1970).

[15] H. Lebesgue, *Measure and the Integral,* ed. K. May (San Francisco: Holden Day, 1966), pp. 13, 17, 81, and 128.

themselves'.[16] This sounds weird. How can there be signs which do not point to anything? How can there be a language waiting for referents? The fact of the matter is that mathematicians do communicate with each other. When Smith tells Jones about his latest discovery in Arkalov Humdrum Theory, Jones understands Smith's *meaning*. Of course, there are some 'nonstandard interpretations' of Arkalov Humdrum Theory, but, in actual practice, this does not prevent Jones from 'seeing what Smith has in mind'.

A third difficulty for formalism is consistency. If a branch of mathematics describes an independently existing reality, then there is no question that it is consistent. However, if a branch of mathematics is merely an arbitrary symbol game, then how can one be sure that, given its rules, it is consistent? Hilbert originally thought that one might be able to prove the consistency of certain branches of mathematics, but, in 1931, K. Gödel showed that, for branches of mathematics which include arithmetic, this is not possible. On a formalist understanding of mathematics, there is no way of ruling out the possibility that, some day, someone will give a valid arithmetical proof of the assertion $0 = 1$.

4.10 Platonism

Realist metaphysicians not drawn to materialism may wish to consider a possibility raised by Plato:

(L) *Platonism:* There are mathematical objects, such as the natural numbers, which exist independently of human minds, and all these objects are immaterial objects.

According to Platonism,

> Even if there were no human being to think about it, there would still be a prime number between 10 and 20.

Note that the Platonist does not have to hold that every possible mathematical object exists in the Platonic empyrean. It may be that there are fractions, but no inaccessible cardinals. It may be that, although Euclidean triangles exist independently of the human mind, hyperbolic triangles do not.

G. H. Hardy (1877–1947) was a Platonist. He compared mathematical objects to mountains, and the communication of mathematical ideas to talk between two climbers who are trying to discuss a distant peak. 'You see

---

[16] D. Hilbert, 'On the Infinite', p. 192.

the reddish cliff? Now look along the ridge on the right, past that cloud. Do you see the twin peaks?'[17]

Another Platonist was K. Gödel. He asserted that the assumption that there are mathematical objects is

> quite as legitimate as the assumption of physical bodies and there is quite as much reason to believe in their existence. They are in the same sense necessary to obtain a satisfactory system of mathematics as physical bodies are necessary for a satisfactory theory of our sense perceptions.[18]

Gödel also wrote:

> despite their remoteness from sense experience, we do have something like a perception also of the objects of set theory, as is seen from the fact that the axioms force themselves upon us as being true.[19]

For Gödel, the interaction between human minds and sets is no more or less mysterious than the interaction between human minds and physical objects. For Gödel, mathematical insight is like a sixth sense which observes mathematical entities.

What are the advantages of Platonism? To start with, Platonism enjoys the advantages of realism. For example, mathematical existence statements do not have to be 'reduced' to something else. For the Platonist, the set of unicorns is empty simply because there are no unicorns, while the set of even primes is nonempty simply because there is an even prime. Furthermore, if Platonism is true, there is no particular need to worry about the use of nonconstructive methods, impredicative definitions, or the Law of the Excluded Middle. Nor is there any need to worry about the consistency of arithmetic, since the natural numbers already have a 'model' in the real world. Platonism, moreover, can make perfect sense of the feeling that some mathematical objects are, not created by us, but discovered. Finally, Platonism has an advantage not shared by other forms of realism: it is perfectly open to the infinite. As M. Resnik writes in his article 'A Naturalized Epistemology for a Platonist Mathematical Ontology':

> Platonism succeeds because, unlike nominalism, materialism and constructivism, it can supply the vast infinities of objects that mathematics requires — more objects than any [finite]

---

[17]G. H. Hardy, 'Mathematical Proof', *Mind* 38 (1929), 18.

[18]K. Gödel, 'Russell's Mathematical Logic', in *Philosophy of Mathematics*, pp. 456–7.

[19]K. Gödel, 'What is Cantor's Continuum Problem?', in *Philosophy of Mathematics*, pp. 483–4.

mind or minds could construct, more objects than the physical universe contains.[20]

Of course, Platonism is not without weaknesses. One weakness is its depiction of mathematics as alien and inhuman. The mathematician is portrayed as a Prometheus ascending to The Realm of the Abstract. In actual fact, the mathematician is often a pretty ordinary person who blunders into an almost inevitable extension of previously known theorems.

A second difficulty for Platonism is the problem of 'multiple reference'. Suppose Smith says

I like the real numbers.

Since the real number system is a complete ordered field, Smith is referring to at least one of these many Platonic objects, but to which one? Each complete ordered field is structurally identical to any other, but each can be seen to contain another one which is, in some sense, more complicated. Is Smith referring to the simplest, most beautiful complete ordered field? Or to the complete ordered field described in Birkhoff and Mac Lane's *A Survey of Modern Algebra*? Or to the class of all complete ordered fields? Or to a complete ordered field which is the only complete ordered field which Smith happens to be observing at that moment — with some mysterious sixth sense? If Platonism is true, then mathematical language refers to independently existing entities, but in a problematic manner.[21]

In any case, as Maddy points out, metaphysicians attracted by Platonism have a choice between the following versions:

(M)　　*Interactive Platonism:* Platonism is true, and human beings can interact with immaterial mathematical objects which exist independently of human minds, in some way 'perceiving' those mathematical objects.

(N)　　*Isolationist Platonism:* Platonism is true, but human beings cannot interact with immaterial mathematical objects which exist independently of human minds.[22]

Examples of Interactive Platonists are Plato himself, K. Gödel, and J. R. Brown. The relevant interaction is observed in (1) the 'felt truth' of

---

[20] M. Resnik, 'A Naturalized Epistemology for a Platonist Mathematical Ontology', in *Mathematical Objects and Mathematical Knowledge,* ed. M. Resnik (Aldershot: Dartmouth, 1995), p. 475.

[21] See M. Dummett, *Truth and Other Enigmas* (Cambridge: Harvard University Press, 1978), p. 207 ff. The 'multiple reference' objection to Platonism is tackled in Chapter 12 below.

[22] P. Maddy, *Realism in Mathematics* (Oxford: Clarendon Press, 1990), pp. 28-35.

certain basic principles, or (2) the 'flash of insight' experienced by someone making a mathematical discovery, or (3) the recognition of a pattern in some physical state of affairs.

On 'felt truth', Brown has the following to say.

> If there were no abstract objects, then we wouldn't have intuitions concerning them; '2 + 2 = 4' would not seem intuitively obvious. It is the same with teacups; if they did not exist I wouldn't see any.[23]

As for the 'flash of insight', one can cite the exclamations of students:

> Oh! *That's* what a function is. I see.

> Suddenly, I saw *why* there are two solutions.

Concerning pattern recognition, take the example of someone studying the scattering of bullet holes around the centre of a shooting target. After a moment of reflection, she says

> That's due to a two dimensional normal distribution.

From the point of view of Interactive Platonism, her identification of the pattern can be interpreted as a mark of contact with timelessly existing mathematical objects.

Sceptics of Interactive Platonism ask *how* the interaction occurs. To this one might reply that we do not doubt the existence of physical objects on the grounds that we cannot explain the link between them and the mind, and so we should not doubt the existence of abstract objects because we cannot explain the link between them and the mind.[24]

Furthermore, abstract objects are abstract, not because we cannot interact with them, but because they are nonphysical patterns or because they are equivalence classes. Such things do not, of course, perturb gravitational fields, but it is common experience that they perturb our minds. Consider the Chinese language, for example. This is an abstract object, but we none the less interact with it when we learn it by means of listening to tokens of the types which are its words.

Adherents of Isolationist Platonism include W. Quine and H. Putnam. On their view, mathematical objects exist, independently of us, because they are essential components of the 'best' theory of the world. They

---

[23] J. R. Brown, *The Laboratory of the Mind* (London: Routledge, 1991), p. 64.

[24] Perhaps Carl Jung's acausal 'synchronicity' can provide a key for understanding how we 'intuit' abstract objects.

are 'ideal objects', and, as such, they cannot be 'perceived'.[25] Isolationist
Platonism has the advantage that its adherents do not have to explain the
'interaction', but it has the disadvantage that it is more vulnerable to the
problem of 'multiple reference', since it portrays mathematical objects in
such a way that we have an essentially indirect access to them.

Isolationist Platonism is defended by B. Hale in his 1994 *Philosophical
Review* article 'Is Platonism Epistemologically Bankrupt?'

Finally, there is a choice between the following:

(O)   *Theistic Platonism:* Platonism is true, and the immaterial mathematical objects which exist independently of the human mind exist only
because a God causes them to exist (for example, as the contents of certain
ideas in God's mind).

(P)   *God-neutral Platonism:* Platonism is true, and the immaterial mathematical objects which exist independently of the human mind also exist
independently of any divine mind.

As we noted in Chapter 1, Theistic Platonism was introduced by a neo-
Pythagorean, Nicomachus of Gerasa (100 AD). It was later taken up by
Augustine.

One difficulty for Theistic Platonism is the possibility that God is so far
above mathematics that it does not even enter his mind. To this difficulty
one can reply that, since God is omniscient, he presumably knows and
understands facts such as the fact that

> A typical human would know that if 4 boxes contain 6 marbles
> then at least one box contains at least 2 marbles.

Note that if Theistic Platonism is correct, there are two ways in which
a mathematical object can exist independently of human minds. First, it
can exist 'in God's mind', as something he thinks about, and, second, it
can exist as something God somehow creates. For example, there is set
theory with 'huge' cardinals and set theory with no huge cardinals. Both
structures exist in the mind of God, in the sense that he is fully aware of
both of them. However, there is still the possibility that God has freely
chosen to create, in some sense, a collection of sets which, as a contingent
fact, include huge cardinals.

Theistic Platonism allows arguments of the following sort:

---

[25] See P. Maddy, *Realism in Mathematics,* pp. 28-35, and also C. S. Chihara, *Constructibility and Mathematical Existence* (Oxford: Clarendon Press, 1990), pp. 9-10.

Huge cardinals are neither necessary nor beautiful.
If a mathematical object is neither necessary nor beautiful then
God does not create it, and it does not exist.
Therefore there are no huge cardinals.

It also allows arguments such as this:

If there are no infinite sets, then it is wrong to praise God for
being able to love us for infinitely many years.
But it is not wrong to praise God for this.
Hence there are infinite sets.[26]

Incidentally, I do not know of any Isolationist Theistic Platonists: most
theists believe that God can communicate some of his mathematical ideas
to human beings, and, in this sense at least, human beings can interact
with those ideas, and hence, at least indirectly, with the immaterial mathematical objects themselves.

4.11 Summary

We have described the various ontological schools in the philosophy of mathematics. In particular, we have described those very important schools
called 'intuitionism', 'formalism' and 'Platonism'. We noted that each
school has its strengths and weaknesses. We also noted that intuitionism
and formalism have trouble accommodating infinite sets, leaving Platonism
as the only major doctrine which is fully open to the infinite.

The 'taxonomy' given in this chapter is summarised, à la Plato, by the
following 'dialectical diagram'.

---

[26] Note that to say that Theistic Platonism 'allows' these arguments is not to say that
it endorses their premisses. It is merely to say that it is willing to take such arguments
seriously, not ruling them out immediately, on the basis of some prejudice against religion.

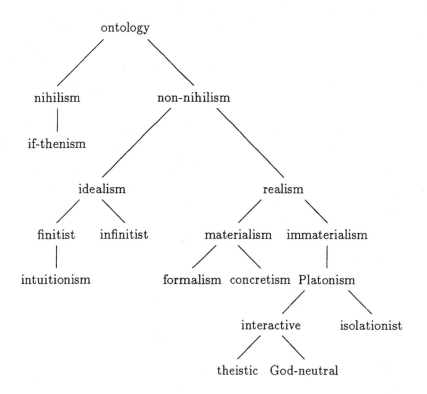

A Taxonomy of Ontology for Mathematical Objects

4.12 Logicism

Logicism is the view that mathematics is a branch of logic. The reason we did not mention logicism in this chapter is that it is not a thesis about the ontology of mathematical objects so much as a view about the relationship between mathematical objects and logical objects, and one that is compatible with various different ontologies. As A. Robinson says, 'the tenets of Logicism seem to be compatible with diametrically opposed views on the problem of existence in Mathematics.'[27]

4.13 Structuralism

Structuralism is the view that mathematics is primarily concerned with abstract structures. For example, it is the abstract group which is important, rather than its various concrete instantiations. Indeed, if two concrete groups are 'isomorphic' — instantiating the same abstract group — we ought to ignore the nonstructural differences between them, and 'identify' them.

Category theory incorporates the structuralist attitude. Its 'objects' are structures, and these structures are defined only 'up to isomorphism'. The number 2, for example, is understood only in terms of a position in a structure called a 'natural numbers object'.

As noted above, structuralism is not an ontological view. Nothing is said about whether the structures are fictions or mental entities or Platonic forms. What structuralism does say is that the correct question is not whether, say, the number 2 exists as the set $\{\{\phi\}\}$ and not otherwise, but whether some natural number structure containing the number 2 exists.

Structuralism is not unproblematic. Its defenders need to answer questions such as the following.

1. Why are we entitled to identify merely isomorphic objects? Is this not just a lack of rigour?

2. Do structures have structure? Can a structure have itself as its own structure?

3. Can an infinite collection have a finite structure, so that, in identifying the collection with its structure, we are, in effect, identifying the infinite with the finite? Is such an identification legitimate?

---

[27] A. Robinson, 'Formalism 64', in *Logic, Methodology and Philosophy of Science*, p. 228. As for the truth of logicism, it depends on what you mean by 'logic'. Mathematics is not a branch of first order logic, but it is a branch of second order logic (with infinite domains). Of course, if you think that infinite domains are not part of logic, you cannot be a logicist — unless, perhaps, you are willing to understand a typical theorem *p* to mean something like 'if the Axiom of Infinity is true then *p*'.

4. Do structures exist independently of the objects whose structures they are? Does it make sense to have just structures without also having some less abstract entities whose structures they are?

5. Would the axioms of set theory make sense if set theory was mere structure? For example, why would one accept the power set or foundation axiom, unless one took seriously the sets themselves and not just their structure?

6. Does the Law of the Excluded Middle apply to propositions about structures, and, if not, can they be real?

7. What is the ontological status of the structures in mathematics?

Chapter 5

Mathematics and Truth

In this chapter, we take up the topic of truth in mathematics. We look into an alleged 'loss of truth', and we discover that even proponents of 'absolute truth' can still find plenty of truth in mathematics.

5.1 Did Truth Get Lost?

Prior to the nineteenth century, mathematicians had thought of the axioms of their subject as necessary, absolute, self-evident, a priori *truths.* The mere possibility of alternative algebras or geometries was unthinkable. In the second edition of his *Nouveaux Elemens des Mathématiques,* published in 1689, Jean Prestet defines an axiom as 'a proposition of which the knowledge is so clear that one cannot doubt it. . . . any reasonable man is forced to agree with it.'[1] From about 1830 on, however, mathematicians began to study geometries incompatible with the standard Euclidean geometry, and some concluded that mathematical axioms, far from being 'absolutely true', were merely arbitrary human conceptions, related to reality (whether material or immaterial) only accidentally. Historian Morris Kline records the history of this shift under the title 'The Loss of Truth'.[2] Indeed, some have gone so far as to conclude that, because there is no 'absolute truth' in geometry, there is no 'absolute truth' whatsoever. Eli Maor, for example, calls his chapter on nineteenth century geometry 'The Vain Search for Absolute Truth'.[3] Yes, a rumour has gone out that there is no truth in

---

[1] Jean Prestet, *Nouveaux Elemens des Mathématiques,* 2nd edn., 1689, p. 6.

[2] M. Kline, *Mathematical Thought from Ancient to Modern Times* (New York: Oxford University Press, 1972), p. 1032.

[3] E. Maor, *To Infinity and Beyond* (Boston: Birkhäuser, 1987).

mathematics. The axioms of Euclidean geometry are at the basis of one structure, and the axioms of 'hyperbolic' geometry are at the basis of another, and neither set of axioms has more claim to truth than the other, and thus, since they contradict each other, neither has any claim at all. Even a statement like

> All squares are quadrilaterals

is not a truth — for who says there are any squares? Perhaps the only squares are physical squares, and perhaps physical squares are best described, not as four-sided figures, but as infinite fractals!

5.2 Historical Background

The topic of truth in mathematics is more easily understood if we recall a little history. About 300 BC, Euclid listed some premisses about points and line segments, and proceeded to deduce from them much of the mathematics known at that time — including the laws of arithmetic (expressed in geometric fashion). Euclid did not do this flawlessly. He occasionally used an assumption he had not stated. However, as Hilbert and others have shown, it is possible to make Euclid's deductive edifice perfectly rigorous by adding a few, innocuous axioms to his original list.

Most of Euclid's axioms (both explicit and tacit) are uncontroversial. One is not likely to be suspicious of statements such as

> All right angles are equal [in size] to one another

or

> Things which are equal to the same thing are also equal to one another.

However, there is one initial assumption which is not quite so vapid. This is Euclid's Fifth, or Parallel, Postulate:

> if a straight line falling on two [coplanar] straight lines makes the interior angles on the same side less than two right angles, the two straight lines, if produced indefinitely, meet on that side on which are the angles less than two right angles.[4]

In other words, if C and D are on the same side of AB, and

$$\angle DAB + \angle ABC < 180°$$

---

[4] Euclid, *The Elements,* trans. T. L. Heath (New York: Dover, 1956), vol. 1, p. 155.

then there is (or can be constructed) a point E, on the same side of AB as C and D, where AD meets BC. Euclid's Fifth postulate characterises the 'straightness' of straight lines in such a way that AD cannot approach BC the way a curved line can approach an asymptote.

For centuries, mathematicians tried to deduce the Fifth Postulate from 'Euclid's other postulates', but failed to do so in a convincing manner. Finally, about 1830, Nikolai Lobachevsky (1793–1856) and others proposed a geometry in which the negation of the Fifth Postulate was added to Euclid's other postulates. This new 'hyperbolic' geometry contains some theorems that might be described as 'hallucinogenic':

H1) Through any point, not on a given straight line, there are infinitely many straight lines each of which is parallel to the given line; hence two straight lines can be parallel without the 'alternate angles' being equal.

H2) If two triangles are similar (that is, have equal angles) then they are congruent; you cannot 'magnify' a triangle without changing its angles.

H3) Some isosceles right triangles have no circumcircles.

H4) There are no quadrilaterals with four right angles; hence the plane cannot be represented by means of the usual graph paper, which divides the plane into squares.

H5) There is a circle greater in area than any triangle.

H6) The Theorem of Pythagoras fails to be true, and hence the distance formula (which is central in analytic geometry and complex analysis) also fails to be true.

H7) There is a circle such that the area of an inscribed regular hexagon is less than that of one of the six segments lying in the circle but outside the regular hexagon.

Mathematicians reacted to these hallucinogenic results in two ways:

Reaction 1): Although these results are not logically impossible, they are so weird that we can, and should, conclude that the Parallel Postulate is the only sensible thing to believe in.

Reaction 2): Although these results are weird, they are not logically impossible, so we have discovered a wonderful, new, hallucinogenic geometry, and are no longer bound by Euclidean 'truths'.

In 1868 (twelve years after Lobachevsky's death), E. Beltrami (1835–1900) gave the first proof that hyperbolic geometry is consistent (assuming that Euclidean geometry is). This led people to accept the work of Lobachevsky, and, in the twentieth century, Reaction 2) became a dogma.

5.3 Truth and Meaning

In discussing the 'truth' of a geometry, we need to assume that its primitive terms have a natural or standard interpretation.[5] We need to assume that an expression such as 'straight line' has a meaning or referent. S. Barker points out that some people think there are only two choices.[6]

Choice 1): The term 'straight line' is a meaningless term in an uninterpreted axiom system, and a question such as, 'is there more than one straight line parallel to a given straight line?' can, and should, be given the trivial answer, 'it depends on whether the axiom system implies there is'.

Choice 2): The term 'straight line' refers to something in the physical universe (to, say, light ray paths), and a question such as, 'is there more than one straight line parallel to a given straight line?' can, and should, be given the answer, 'go and ask a physicist; if they do not know, nobody does'.

What Barker wants to say, however, is that there is also

Choice 3): The term 'straight line' refers to an objectively existing, but nonphysical entity, and a question such as, 'is there more than one straight line parallel to a given straight line?' might be given an answer such as, 'consult your mathematical intuition'.

According to Barker, this objectively existing, but nonphysical, straight line might be an abstract set of ordered pairs of numbers, or it might be something more naturally geometrical, something 'more faithful to what we normally *mean* by straightness'.[7]

Kant and Gödel are two philosophers who take Choice 3). For Kant, any human being who considers a pair of parallels cut by a transversal simply 'observes' that

the alternate angles are equal

— a statement which immediately implies the Parallel Postulate.[8]
According to Gödel:

---

[5] A 'primitive term' is a basic undefined term, such as 'point', or 'straight line'.

[6] S. Barker, 'Kant's View of Geometry', in *Kant's Philosophy of Mathematics,* ed. C. J. Posy (Dordrecht: Kluwer Academic Press, 1992), p. 227.

[7] Ibid., p. 233-7, with my emphasis.

[8] I. Kant, *Critique of Pure Reason,* B744-5, and see B39. Aristotle had the same intuition. See *Metaphysics* 1051a21-31.

> In its purely mathematical aspect our Euclidean space intuition
> is perfectly correct, namely, it represents correctly a certain
> structure existing in the realm of mathematical objects.[9]

For Gödel, an axiom of geometry, interpreted in a natural way, is true just in case it is a correct description of an actual state of affairs. These states of affairs have a causal effect on us in the sense that we can somehow 'intuit' them.

For example, I can 'intuit' a grid of squares, such as one finds represented on a piece of graph paper, and thus I can 'intuit' the existence of squares. From this I can conclude that the assumption 'there are squares' is true, and, knowing that hyperbolic geometry implies that there are no squares, I can conclude that hyperbolic geometry is false. Indeed, I can conclude this more directly from my intuitions about straight lines: if two straight lines approach each other, then, of course, they meet.

5.4 What is Truth?

In this section we examine six major philosophical positions on the nature of truth, to see what each of them implies for the 'truth' of Euclidean geometry. What we shall discover, perhaps surprisingly, is not a 'loss of truth', but a 'feast of truth'.

Note that each of the six positions is a metaphysical position on the essence or definition of truth — rather than, say, an account of how truth is recognised, or a view about how ordinary language speakers use the phrase 'is true'.

Coherentism

The coherence theory of truth is the view that a proposition is true if it is consistent with 'given' knowledge. For example, the law $e = mc^2$ is true in the coherentist sense if it is consistent with various other laws of physics, and with various statements recording experimental observations. Furthermore, each of these other laws of physics, and each of these observation statements is true, in this sense, if it is consistent with other laws of physics, and other observation statements. The coherentist is willing to live with the implied circularity, and with the fact that the set of statements obtained by negating a set of 'truths' is not less consistent than the original set of 'truths' is.

---

[9]M. J. Greenberg, *Euclidean and Non-Euclidean Geometries* (San Francisco: W. H. Freeman, 1974), p. 260.

Let A denote the conjunction of Euclid's axioms (explicit or implicit), with the exception of the Parallel Postulate. Let PP denote the Parallel Postulate itself. If the 'given knowledge' consists of A — this under some standard or even non-standard interpretation — then each of PP and its negation are 'true', under the same interpretation, since, precisely, both the Parallel Postulate and its negation are compatible with Euclid's other postulates. So, in this case, we have a paradox: PP is true and not-PP is true. On the coherence theory of truth, we must accept Euclidean and nonEuclidean geometry on equal terms — and simply endure the inconsistency.

Pragmatism

The pragmatist theory of truth is the view that a proposition is true if it is on the whole, in some sense, advantageous to believe it. On this view, Euclidean geometry is sometimes true and sometimes false. In the time of Euclid, it was advantageous to believe it, since that belief helped focus the research efforts of ancient Greek mathematicians in their investigation of basic mathematical properties. In the time of Lambert, however, when much of the basic mathematics had already been worked out, it was not advantageous to believe in Euclidean geometry, since that belief retarded the development of non-Euclidean geometry.

Suppose, moreover, that, under some natural interpretation, three-dimensional hyperbolic geometry is an exact description of physical space. Suppose also that, under the same interpretation, three-dimensional Euclidean geometry is such a good approximation to physical space that it is almost impossible to measure the empirical data accurately enough to detect the discrepancy.[10] Suppose, finally, that it is on the whole more advantageous to believe that space is Euclidean — for one thing, it makes it easier to teach or learn physics. Then, on the pragmatist view, it is 'true' that space is Euclidean.

Relativism

Relativism is the view that a proposition is 'true' if and only if it is 'true for some person X'. Moreover, 'p is true for X' just in case X believes that p (or is firmly convinced that p, or some such thing). As Protagoras phrased it, any given thing

---

[10] For all we know, this may be the actual situation. See R. Penrose, 'The Geometry of the Universe', in *Mathematics Today,* ed. L. A. Steen (New York: Springer-Verlag, 1978), p. 91.

is to me such as it appears to me, and is to you such as it appears to you.[11]

On this view, Euclidean geometry is true for supporters of Euclidean geometry, while hyperbolic geometry is true for supporters of hyperbolic geometry. Indeed, the statement

Geometric points do not hold pints

is true for those who believe that geometric points do not have any volume, but the statement

Geometric points do hold pints

is true for those who believe that Hilbert should have taken things even further. Even a statement such as

Triangles have four angles

is no more or less true than any other statement — for those for whom it is true.[12]

Verificationism

Verificationism is the view that a statement is true just in case it can be known that it is the case, using 'scientific' methods. The word 'scientific' is to be understood here in the context of British empiricism. 'Scientific' methods include empirical observation, mathematical deduction, and the experimental testing of hypotheses. Examples of statements open to verification are

That stick is approximately 1.23 metres long

or

$$3 + 2 = 5$$

or

Any two bodies attract each other on account of their mass.

'Scientific' methods do not include perfectly exact measurement or infinite calculations.[13] An example of a statement which is not open to 'scientific' verification — and *a fortiori* not 'true' (in this sense) — is

---

[11] Plato, *Theaetetus* 152a. In the classification given by R. L. Kirkham in *Theories of Truth* (London: MIT Press, 1992), what we call 'relativism' is labelled 'subjectivism'.

[12] As Plato pointed out to Protagoras (in *Theaetetus* 169a), mathematics contains certain 'core' truths, and anyone who denied that, say, $2 + 2 = 4$ would simply be wrong.

[13] J. S. Mill, *Autobiography*, ed. H. J. Laski (New York, 1952), pp. 225-6.

Physical space is perfectly Euclidean.

For the verificationist, sense perception is an acceptable 'scientific' method, but spiritual discernment is not. There is a materialist bias in the verificationist theory of truth, and the verificationist faces a special problem in dealing with the putatively immaterial objects of mathematics. How can the verificationist claim that an axiom of mathematics, interpreted in an abstract fashion, is 'true', if he cannot empirically observe its truth, or deduce its truth from the results of an empirical observation?

A standard verificationist solution to this problem is the 'theoretical construct' solution: a mathematical object exists, or a set of mathematical axioms is true, if the object or set of axioms plays an essential role in any reasonably simple, empirically sound theory of the physical universe.[14] For example, the axioms of the real number system play an essential role in any reasonably simple, empirically sound theory of motion, and therefore we may take it that the axioms of the real number system are 'scientifically' true.

According to the verificationist view, then, it may be the case that several incompatible geometries are true. Euclidean geometry is true because it underlies, in an essential way, the complex analysis that goes into many explanations in physics. Hyperbolic geometry is also true because it underlies, in an essential way, some of the cosmological modelling that is used in astronomy.

Assertabilism

Assertabilism is the view that a statement is true if and only if there is a finite procedure which results in evidence which confirms that statement. For example, the statement

There is a prime between a million and two million

is true because there is a calculation which warrants asserting it.[15]   M. Dummett sometimes writes like an assertabilist:

> We are entitled to say that a statement P must be either true or false, that there must be something in virtue of which either it is true or it is false, only when P is a statement of such a kind that we could in a finite time bring ourselves into a position in

---

[14] A. J. Ayer, 'The *a priori*', in *Philosophy of Mathematics*, 2nd edn., pp. 315-28.

[15] C. J. Posy, 'Kant's Mathematical Realism', in *Kant's Philosophy of Mathematics*, ed. C. J. Posy (Dordrecht: Kluwer Academic, 1992), pp. 299-304.

which we were justified either in asserting or in denying P; that is, when P is an effectively decidable statement.[16]

For example, assume that we have no way of knowing what occurred before the Big Bang. Let P be the statement

Before the Big Bang there was a city of five million inhabitants somewhere in the universe

Given our assumption, neither P nor not-P is assertable, and hence neither P nor not-P is true.[17]

As for mathematics, if the evidence in question is limited to 'scientific' evidence, then assertabilism is a close cousin of verificationism, and it may well be possible to conclude that both Euclidean and nonEuclidean geometry are 'true' in the assertabilist sense.

Realism

By *realism*, in this context, we mean the correspondence theory of truth. This is the view that a statement is true just in case it correctly describes an actual state of affairs. Aristotle puts in this way: 'to say of what is that it is, and of what is not that it is not, is true'.[18]

K. Popper was a correspondence theorist. Drawing on the work of A. Tarski (1901–1983), he explained the correspondence theory as follows. In order to discuss the 'truth' or 'correspondence' of a statement P in a language L, we use a meta-language L'. The meta-language has names, like 'P', for the statements of L, and also names, like 'p', for descriptions in L' of states of affairs. For example, if L is German, and L' English, then P might be

Der Mond ist aus grünen Käse gemacht

and p might be

The moon is made of green cheese.

Here P *corresponds* to the facts (and hence is *true*) just in case p. In general, if p is an expression in L' which describes the (supposed) fact which P (in L) describes, then P *corresponds* to the facts (and is *true*) if and only if p.[19]

---

[16] M. Dummett, 'Truth', in *Truth,* ed. G. Pitcher (Englewood Cliffs: Prentice-Hall, 1964), pp. 108-109.

[17] Ibid., p. 108.

[18] Aristotle, *Metaphysics* IV 7.

[19] K. R. Popper, *Objective Knowledge* (Oxford: Clarendon Press, 1979), pp. 314-6.

According to the correspondence theory, Euclidean geometry, under some interpretation of 'point', 'straight line', and so on, is true if it is a correct description of some actual state of affairs.[20]

The meta-language L' is introduced partly to deal with the 'Liar Paradox':

> Let P be the statement
>
> > This statement is false.
>
> If P is true then it is false and vice versa.

The resolution of the Liar Paradox goes as follows: If P is a statement, then it belongs to some language L, and it describes some state of affairs s. However, what is the state of affairs s which P describes? It has to do with something (the falsity of a statement in L) which cannot be described without using a meta-language L'. Hence the statement P actually belongs to the meta-language L', rather than to the language L. The assumption that P is a statement thus leads us to the contradiction that, although P is in L, actually P is not in L. Hence P is not a statement, and there is no Liar Paradox.

Now suppose there is an immaterial geometric structure, existing independently of human beings. It might, for example, exist in the mind of God. There are several possibilities.

(A) The structure is correctly described by a natural interpretation of A-and-PP (Euclid's full system), and hence the Parallel Postulate corresponds to something in reality.

(B) The structure is correctly described by a natural interpretation of A-and-not-PP, and hence the negation of the Parallel Postulate corresponds to something in reality.

(C) The structure is a set-theoretical-equivalence-class-Dedekind-cut version of the real number system. It is geometric only in the sense that it contains numerical 'models' for different kinds of geometry (providing ordered pairs of real numbers for the 'points', and special sets of ordered pairs of real numbers for the 'lines'). There is a 'model' of Euclidean geometry in it, and there is also a 'model' of hyperbolic geometry. Thus both geometries

---

[20]Note that if, under some interpretation, Euclidean geometry is true in this sense, then, under a different (counter-intuitive) interpretation, hyperbolic geometry is also true in this sense. This is because Euclidean geometry contains the Poincaré model for hyperbolic geometry.

are 'true', but only if they are interpreted as indirect ways of talking about real numbers. For example, theorems in Euclidean geometry about circles are 'true' in the sense that they accurately describe sets of the form

$$S = \{(x,y) \in \mathbf{R} \times \mathbf{R} \mid (x-h)^2 + (y-k)^2 = r^2\}$$

(D) Possibility (A) is true, and, moreover, Possibility (B) is false: there is no natural interpretation of 'straight line' according to which two straight lines can approach each other indefinitely and not meet.

For the correspondence theorist who accepts mathematical states of affairs and accepts, moreover, 'standard' interpretations of mathematical formulas, each of the above possibilities is a possibility of having truth.

However, what if the correspondence theorist rejects mathematical states of affairs, or rejects 'standard' interpretations? In his book *Focusing on Truth,* L. E. Johnson takes a view of mathematics according to which there are no 'platonic-style entities', and mathematical statements *per se* are uninterpreted. Given Johnson's sympathies for the correspondence theory of truth, the result is not surprising. He writes:

> mathematical 'propositions' are not things of the sort which can be true or false.[21]

For Johnson a statement such as

$$\int_0^1 \sin x \; dx = \int_0^1 \sin x \; dx$$

is not a truth about 'platonic-style areas'. Nor is it a truth about anything else. It is not a truth at all. Indeed, Johnson would agree with Vaihinger that no mathematical statement whatsoever is true. There is, not just a 'loss of truth', but a total annihilation of truth.

Johnson's position would be of interest to certain students seeking a panacea for their true-false tests, but, given its extremity, it would not be of much interest to working mathematicians. I suggest we classify it as a curiosity.

Summary

According to each of the above theories of truth, it is quite possible that Euclidean and nonEuclidean geometry are both true. In particular, on the correspondence theory of truth, it is quite possible that both geometries are

---

[21] L. E. Johnson, *Focusing on Truth* (London: Routledge, 1992), p. 226.

'absolutely true' — about different things.[22] Moreover, it is also possible, on the correspondence theory of truth, that one geometry is 'absolutely true' (describing the actual state of affairs in the Platonic realm), while the other is 'simply false' (having to be interpreted in some wholly unnatural or nonstandard way in order just to have a model). The fact that there are two 'equally consistent' geometries may imply a loss of simplicity, but it certainly does not imply any 'loss of truth'. Indeed, you can take whatever theory of truth you like, and there is still plenty of room for truth in mathematics.

5.5 Language Fuzziness

Certain anti-truth philosophers of mathematics claim that one ought to be bewildered by the vast number of different possible interpretations, or 'models', for any given mathematical structure.[23] Let L be a list of statements expressing all our intuitions about, say, real numbers. Let A' be an axiomatic presentation of the real numbers. Now A', they point out, is open to an infinite number of 'non-standard' interpretations, while L is finite. Hence L cannot be very useful for telling us what real numbers are 'really like'. You can't hold sand in a sieve.

This argument is undermined by at least three objections.

First, there is no a priori reason for thinking the human mind is finite, so that the list L must be finite too.

Second, even if our intuitions about real numbers are finite in number, they may still be capable of rejecting, in a reliable fashion, all but a finite number of models as false. For example, in many cases, our intuitions are quite capable of telling us, in a reliable fashion, that a model is too complicated to be a 'true model' of the reals. Since the number of 'sufficiently simple' models is finite, there is no reason to think that a human being cannot say precisely what a 'real number' is.

Third, mathematicians succeed in communicating their ideas. It may be true that their words can be given infinitely many different meanings, but the fact is that the second mathematician understands the first. When the second mathematician exclaims, 'Now I see!', then he or she is usually referring to just what the first mathematician is referring to. In actual practice, human intuition is not bogged down by the sceptic's multiple reference problem. Human intuition is more powerful than the anti-truth

---

[22] People use the term 'absolute truth' to refer to that hard, objective stuff which is actually exactly what correspondence theorists have in mind.

[23] See M. Dummett, *Truth and Other Enigmas* (Cambridge: Harvard University Press, 1978), p. 207 ff.

philosopher of mathematics is willing to allow.[24]

5.6 Revolutions in Mathematics

There has been some discussion about 'revolutions' in mathematics.[25] This discussion originated with the work of T. Kuhn, who maintained that science, instead of being cumulative, sometimes overthrows established 'truths', and replaces them with new 'paradigms'. M. Crowe claimed that nothing quite so radical occurs in mathematics, and yet, as he later admitted, one can give a reasonable definition of 'revolution' according to which there are revolutions even in mathematics.[26]

In closing this chapter, we consider whether the revolutions in mathematics give us any special reason for thinking there is no 'absolute' truth in mathematics.

Let us begin by listing a few examples of statements once thought to be true, but now discarded:

Every distance is rational (Pythagoras)

There are no infinite collections (Aristotle)

There is no square root of -1 (Cardano)

Every property determines a set (Frege)

There are only 8 essentially different ways of tiling the plane with congruent, convex pentagons (Kershner)

Each of the above statements was, at one time, taken as 'axiomatic' or 'proven', but was later rejected, either because it met with a counter-example, or because the progress of mathematics insisted on a wholly different point of view. This historical fact may lead one to wonder if even a truth like '$2 + 2 = 4$' might someday be abandoned, and it may also lead one to wonder whether, someday, *all* our current mathematics might be considered obsolete.

This kind of wondering is not realistic. Whether in mathematics or science, no belief is overthrown unless other beliefs are retained. This is

---

[24]Note that, just as my inability to explain sense perception does not entail that I cannot talk to you about trees, or decide whether something is a tree, so my inability to explain mathematical intuition does not entail that I cannot talk to you about real numbers, or decide that this beer mug is not the number $-\pi$.

[25]D. Gillies, ed., *Revolutions in Mathematics* (Oxford: Clarendon Press, 1992).

[26]M. Crowe, 'Ten "Laws" Concerning Patterns of Change in the History of Mathematics', and 'Afterword', in *Revolutions in Mathematics*.

because we need reasons in order to discard principles, and those reasons have to come out of an accepted methodology or knowledge. Historians are free to emphasise the revolutionary aspects of science, if they wish, but it would be equally valid to emphasise the constant aspects of science. For example, there has never been a revolution in science which discarded the principle

> Inconsistency is to be avoided

or the principle

> If a law predicts that you will observe p in circumstance c, and you never observe p in circumstance c, then the law is not correct.

Indeed, one could not imagine a revolution in which principles such as these were discarded, since, precisely, it is only in terms of principles such as these that *any* principles can be discarded. Pro-revolution historians look for laws that have been overthrown. They hint that no law is ever entirely safe, and they may even fantasise that every law will eventually be discarded. Non-revolution historians point out that certain basic methodologies and certain basic facts remain, and must remain, constant. There will never be a revolution which rejects simple arithmetic, or overthrows the fact that water is less dense than gold. If, tomorrow, someone finds an odd perfect number, there is no particular reason to think that, in hundred years, number theory textbooks will offer a 'proof' of the 'theorem' that there are no odd perfect numbers. Given the fallibility of human beings, there are no doubt a few currently accepted 'theorems' which will have to be rejected, but this by no means implies that someday *all* our beliefs will be overthrown. It turned out that, an alleged proof notwithstanding, it was not 'absolutely true' that there are only eight different ways of tiling the plane with congruent, convex pentagons. But this does not imply that '$2 + 2 = 4$' is in danger, and it does not imply that there are no absolute truths whatsoever, in any branch of knowledge. Compare the statements

> We occasionally think that something is true, when it is not

and

> Nothing we think is true is really true.

The fact that we are prone to error implies the first statement, but it certainly does not imply the stronger second statement. All-or-nothing thinking tends to blur the distinction between the two statements. Philosophy, let us hope, sharpens it.

5.7 Appendix: Tarski on Truth

One of the reasons the metalanguage is introduced in connection with the correspondence theory of truth is that if L is a consistent language including arithmetic, the predicate 'true in L' is not a predicate in L, but only in some extension of L. This was discovered by Alfred Tarski in 1931. We can give an informal version of his proof as follows.

Let f be a bijection between the wff's (well-formed formulas) of L and $\mathbf{N}$. Let $s : \mathbf{N} \longrightarrow \mathbf{N}$ such that $s(x) = 0$ if $f^{-1}(x)$ is *not* a wff with exactly 1 free variable, and $s(x) = f(f^{-1}(x)|_x)$ otherwise. (In the 'otherwise' case, $s(x)$ is the number f associates with the wff obtained by substituting x for the unique free variable in $f^{-1}(x)$.)

To obtain a contradiction, suppose L has a 'truth predicate' P such that, for any closed well-formed formula q in L (one with no free variables), we have $Pf(q)$ iff q.

Note that *not* $Ps(x)$ is a wff with exactly 1 free variable, namely, x. Let

$$t = f(not\ Ps(x))$$

Then t is a constant, and *not* $Ps(t)$ is a wff with no free variable. Moreover,

$$s(t) = f(f^{-1}(t)|_t) = f(not\ Ps(x)|_t) = f(not\ Ps(t))$$

and hence

$$Ps(t)\ \text{iff}\ Pf(not\ Ps(t))$$

But the formula after the 'iff' is equivalent to *not* $Ps(t)$, since P is a truth predicate. Hence $Ps(t)$ iff *not* $Ps(t)$. Contradiction.

Chapter 6

Mathematics and Values

Films, poems, and paintings all have their critics. In this chapter, we shall be critics of mathematics. We shall try to decide how various factors contribute to making a piece of mathematics better or worse. In 'Some Proposals for Reviving the Philosophy of Mathematics', Reuben Hersch asserts:

> Our very existence as a single profession, and our ability to agree in practice that certain deeds in mathematics are deserving the highest praise and reward, prove that there are common standards of excellence which we use as criteria for evaluating our work. To make these criteria explicit, to bring them into the open for discussion, challenge, and controversy, would be one important philosophical activity for mathematicians.[1]

We shall engage in this important activity here.

Our aim is to develop an axiology for pieces of mathematics. In doing this, (1) we invoke values (such as elegance), (2) we invoke a plurality of values, and (3) we invoke values in a way that is not tied to any particular culture. We begin by saying something about each of these procedures in turn.

(1) There has been some debate about whether values exist, whether there are absolute values, and whether a human being could recognise examples of absolute values, if there were any. Some would claim that a minor generalisation in mathematics is worth less — objectively and absolutely speaking — than an original discovery that solves a long-standing problem.

---

[1] R. Hersch, 'Some Proposals for Reviving the Philosophy of Mathematics', in *New Directions in the Philosophy of Mathematics*, ed. T. Tymoczko (Boston: Birkhäuser, 1986), p. 12.

Others would maintain that value judgements are invariably a matter of individual taste. Such issues need a book of their own. For our purposes here, we shall adopt the view that if a large number of people use a concept to reach conclusions about the worth of something, then it is a meaningful philosophical enterprise to do the same in the case of mathematics.

(2) We shall invoke a plurality of values. Mathematics is rich, and it is a costive philosopher who praises or damns a proof for its correctness, without also attending to its beauty, fruitfulness, or meaningfulness. Mathematics has been described as 'standing on the borderline of all that is beautiful in Art, and all that is wonderful in Science'. Such a vast and lovely domain deserves a critique which takes account of its many qualities, which can tolerate the absence of some virtues as long as others are present. A solution can be 'good' even if it contains a logical gap — just as it can be 'bad' in spite of impeccable rigour.

(3) An axiology for mathematics should not be tied to any one culture. Mathematics is an international (if not interplanetary) activity, and it needs a universal axiology. There are, to be sure, some pieces of mathematics which are 'good' relative to some cultures but not to others. Euclid's geometric derivations of algebraic identities might be a case in point. We now find them clumsy, but the Greeks placed a higher value on geometric methods than we do. On the other hand, there are many pieces of mathematics which almost everyone will agree about. If a schoolboy turns in a repetitious, ten page derivation of the formula for the roots of a quadratic polynomial, no one in any culture is going to applaud. Again, just as almost all writers acknowledge the genius of Shakespeare, so almost every mathematician values Euclid's proof that no finite list of primes contains every prime.[2]

In *The Concepts of Value*, Aschenbrenner looks at more than a thousand appraisive terms.[3] Many of them might be used in evaluating mathematical proofs. We shall concentrate on just twenty-four terms, occurring in twelve pairs of 'opposites'. As we shall show, these pairs provide a practical structure for judging works of mathematics. These 'opposites' are the following:

complete–incomplete
rigorous–intuitive
elementary–advanced
simple–complex
clarifying–mystifying
specific–general

---

[2] Euclid noted that if the list is $2, 3, 5, 7, 11, \ldots, P$, and if N is the product of these primes, then $N + 1$ has a prime factor not on the list.

[3] K. Aschenbrenner, *The Concepts of Value* (Dordrecht: Reidel, 1971).

concrete–abstract
constructive–nonconstructive
practical–impractical
elegant–ugly
meaningful–pointless
fruitful–sterile

We shall now discuss each dichotomy in turn.

6.1 Complete–Incomplete

A mathematical proof can be incomplete in several ways.

(1) It may contain a logical gap which can, however, be filled in without too much trouble. Thanks to the alleged need to 'save space', this occurs in the majority of published articles.

(2) A proof may be incomplete because it overlooks a special case which can, however, be handled in the same way as the rest of the problem. For example, in *Geometry Revisited,* Coxeter and Greitzer show that the three circumcircles of the three equilateral triangles constructed externally on the sides of a given triangle meet in a common point.[4] However, they discuss only the case in which this point lies inside the given triangle. The case in which it lies outside the given triangle is a bit different — although it can be handled in a similar manner.

(3) A proof may be incomplete because it contains a logical gap which no one can fill (or, at least, no one at the present time). Indeed, there are even published 'proofs' of assertions which can be defeated by counterexamples. Perminov does not think this is a serious possibility:

> The passage from individual confidence to that shared by the community does not merely decrease the probability of the proof's deficiency but completely eliminates it.[5]

This is a nice communitarian sentiment, but it is not borne out by the history of mathematics. As Kershner and Wilcox point out:

> Mathematical history contains rare instances of arguments that were generally accepted as proofs for hundreds of years, before

---

[4] H. S. M. Coxeter, and S. L. Greitzer, *Geometry Revisited* (Washington: The Mathematical Association of America, 1967), p. 61. The point in question is called the *Fermat point.*

[5] V. Y. Perminov, 'On the Reliability of Mathematical Proofs', *Revue Internationale de Philosophie,* 42 (1988), 504.

being successfully challenged by a very ingenious mathematician, who pointed out a possibility that had been overlooked in the alleged proof.[6]

Kershner unwittingly provided an interesting example of this. In 1968, in the *American Mathematical Monthly,* he announced that he had found all possible ways of tiling a plane with congruent convex pentagons. He stated that 'the proof that the list ... is complete is extremely laborious and will be given elsewhere'.[7] This proof must have overlooked a possibility, because, in 1975, Richard E. James found a tiling Kershner had missed. Nor was this the end of it. In 1976, a San Diego housewife, Marjorie Rice, decided to make a hobby of pentagon tilings, and succeeded in finding 4 more, all of which had eluded Kershner. The problem is still open.[8]

(4) A piece of mathematics may be incomplete because it depends on a well-known conjecture which has not yet been proved. Some results in computational number theory, for example, are based on the assumption that the Riemann Hypothesis is correct.[9]

(5) A proof may be incomplete because it fails to shed any light on *why* the result it establishes is true. A lucky manipulation of symbols may yield some fact, but without adding to our understanding. For example, consider the Diophantine equation

$$x(x+1)(2x+1) = 6y^2$$

Various solutions have been given for it, but most of them are more mystifying than enlightening. The result falls out by means of a series of tricks or coincidences, and a full understanding of the problem still seems to elude us.[10]

---

[6] R. B. Kershner, and L. R. Wilcox, *The Anatomy of Mathematics* (New York: Ronald Press, 1950), p. 77.

[7] R. B. Kershner, 'On Paving the Plane', *American Mathematical Monthly,* 75 (1968), 840.

[8] D. Schattschneider, 'In Praise of Amateurs', in *The Mathematical Gardner,* ed. D. A. Klarner (Boston: Prindle, Weber & Schmidt, 1981), pp. 140-66. In 1985, yet another tiling was found, by Rolf Stein.

[9] H. Putnam, 'What is Mathematical Truth?' in *New Directions in the Philosophy of Mathematics,* pp. 51-2. If $f(n)$ is the nth prime, $g(n)$ is the product of the first n primes, and $h(n) = 1/(\sqrt{f(n)}(\ln f(n))^2)$, then the Riemann Hypothesis is, in effect, the assumption that there is a constant $C > 0$ such that, for all natural numbers n, $f(n)h(n) - C < (\ln g(n))h(n) < f(n)h(n) + C$.

[10] W. S. Anglin, 'The Square Pyramid Puzzle', *American Mathematical Monthly,* 97 (1990), 120-4, or W. S. Anglin, *The Queen of Mathematics* (Dordrecht: Kluwer, 1995), section 4.4.

We have a certain awe for the so-called certainty of mathematics, and it can be startling to discover a serious gap in an accepted proof. None the less, completeness, I maintain, is only one of several virtues in mathematics. An incomplete proof may put us on the right track. It may be delightful to read. The parts of it which are valid may be highly instructive. The fact that it depends on an unproved conjecture may inspire some mathematician to prove the conjecture. And anyone who believes that incomplete proofs should not be published is out of touch with the historical realities of mathematical research.[11]

6.2 Rigorous–Intuitive

A proof lacks rigour if it relies on intuitions which are hard (or impossible) to defend in a purely logical manner. A paradigm of lack of rigour is Euclid's proof that one can construct equilateral triangles. In Proposition 1 of Book I of the *Elements*, he assumes that a circle with centre A and radius AB meets the circle with centre B and the same radius. The circumferences do not just pass through each other without touching, but actually share a common point. This assumption is justified by the diagram (that is, by our intuitive notion of continuity), but it does not follow from Euclid's explicitly stated postulates.

Another example of an argument lacking rigour is Hao Wang's Paradox:

1 is a small number.
If n is a small number, so is $n + 1$.
Thus every counting number is small.

Here the lack of rigour is due, not to any logical gap, but to the vagueness of the term 'small'.

Rigour is sometimes held up as an ideal. The good mathematician drafts the argument in such a way that each line follows with perfect certainty and clarity from the preceding ones. All hidden assumptions are revealed, and every step is meticulously checked. For, as Russell claimed,

insofar as rigour is lacking in a mathematical proof, the proof is defective.[12]

---

[11] J. V. Grabiner, 'Is Mathematical Truth Time-Dependent?', *American Mathematical Monthly*, 81 (1974), 354-65.

[12] B. Russell, *Introduction to Mathematical Philosophy* (London: Routledge, 1919), p. 144.

There are two problems with this ideal. First, we are fallible, and we sometimes mistake nonrigorous proofs for rigorous ones. Second, too much rigour can slow things down and even impair progress. If one is too fussy about the details, one may miss the significance of guiding intuitions. Let us consider these two problems in turn.

(1) Rigour is not always recognizable

If an assumption is hidden, it is often hard to realise that it is there. If a proof is long, the mathematician's attention may wander, and he or she may confuse two symbols, or make a careless mistake in arithmetic. Human psychology being what it is, we have blind spots, and, as the history of mathematics attests, we sometimes err even when we are careful. Occasionally we conclude that what is actually an invalid proof is valid. We are fallible.

Let us suppose that the probability that a mathematician concludes that an average-length invalid proof is valid is some small number ϵ. From the history of mathematics, we know that $\epsilon > 0$. If 1000 mathematicians independently check the same proof, and all report that it is valid, that greatly decreases the probability of error. It might go down to, say, ϵ^{1000}. On the other hand, if the proof is long, the probability of error will go up. In every case, however, we are dealing, not with certainty, but with probabilities. Human beings are incapable of infallibly recognising lack of rigour.

And so are computers. Our proofs of several theorems, including our proof of the Four Colour Theorem, are based on computer calculations so lengthy that no human could carry them out in a hundred years.[13] The software is 'debugged', and also checked against other software, but it is not verified in a systematic, deductive fashion. Indeed, as De Millo has pointed out, a rigorous verification of computer software is normally not practical: for a typical arithmetical package, the verification would be so lengthy that it would be more open to error than the software it was supposed to vindicate.[14] Of course, it is highly unlikely that a well-tested computer program produce incorrect results, but that is not the point. The point is that rigour is meant to exclude even the smallest chance of error.

In any case, whether we use the computer or not, we do not have the power of distinguishing nonrigorous from rigorous proofs with a perfect

---

[13] Another example is our proof that $x^2 + 28 = y^3$ has only six integer solutions. See W. J. Ellison, F. Ellison, J. Pesek, C. E. Stahl, and D. S. Stall, 'The Diophantine Equation $y^2 + k = x^3$', *Journal of Number Theory*, 4 (1972), 107-17.

[14] R. De Millo, R. Lipton, and A. Perlis, 'Social Processes and Proofs of Theorems and Programs', in *New Directions in the Philosophy of Mathematics*, pp. 268-85.

success rate. The computer may be working perfectly, but we have no practical way of proving this. A proof we are sure is rigorous may actually be so, but we have no infallible way of knowing.

(2) Rigour is not always good

On the view that the primary role of a proof is to secure a result from any possible attack by a sceptic, rigour is paramount. One wants that result with deductive infallibility. There are, however, other views of proof. A proof can serve several other ends: the connection of ideas, the stimulus of the imagination, the expression of beauty, and so on. On a view which does not chain a proof to a single purpose, rigour is merely one of a plurality of considerations.

In 1899, Hilbert produced a rigorous version of Euclid's geometry, in which he gave detailed proofs of many facts that are visually obvious.[15] In the interests of rigour, Hilbert had to take a very slow, meticulous approach, and one might well prefer Euclid's faster, more intuitive presentation. After all, Hilbert does not prove any of Euclid's theorems wrong: he just takes a longer time proving they are right. If Euclid can be faulted for lack of rigour, he should also be praised for a brevity and elegance not found in Hilbert's punctilious checking of betweenness relations.

Rigour can impair progress too. As Kitcher has pointed out, sometimes mathematical advances are made only because rigour is ignored in favour of analogy or extrapolation.[16] Great mathematicians have an intuitive 'feel' for the key to a problem. Sometimes they are wrong, but they are right so often that it is a more efficient use of their time if they leave the rigorous justification of their claims to the students. The development of analysis provides an example. If great mathematicians had stopped to worry about the rigour of, say, the early proofs of the Fundamental Theorem of Algebra, they would not have produced half the powerful results for which we are indebted to them.[17]

When we judge a work of mathematics, then, we should be willing to praise, not only rigour, but also bold intuition — and even employment of empirical data.

---

[15] D. Hilbert, *Foundations of Geometry* (Chicago: Open Court, 1971).

[16] P. Kitcher, *The Nature of Mathematical Knowledge* (New York: Oxford University Press, 1983), p. 240.

[17] See J. Stillwell, *Mathematics and Its History* (New York: Springer-Verlag, 1989), pp. 195-200.

Rigour and History

The relationship between rigour and intuition can be seen in the history of
the idea of continuity. At first continuity was not defined at all. Then, in
1821, Cauchy gave the usual definition of a function's being continuous at
a point: if $f : \mathbf{R} \to \mathbf{R}$ then

> $f(x)$ will be called a *continuous* function, if ... the numerical
> values of the difference
>
> $$f(x + \alpha) - f(x)$$
>
> decrease indefinitely with those of α.

This represented an increase in rigour, but (1) it did not represent any
final stage of rigour, and (2) it may have led us away from the intuitively
correct notion of continuity. As for (1), Cauchy did not say exactly how
the difference is to decrease with α, and he did not say exactly what a
function is. As for (2), Cauchy's definition admits as 'continuous' some
'monsters' which cannot be 'drawn without taking your pencil off the page',
and — at least in its higher dimensional form — it classifies as discontinuous
some functions which are 'all of one piece'. For example, the function
$f(x) = e^{-1/x} \sin(1/x)$ — with $f(0) = 0$ — is continuous, in Cauchy's sense,
at every point. Moreover, all its derivatives are continous. However, if you
put your pencil at the origin, with the intention of drawing this function
over the interval from $x = 0$ to $x = 1$, you will find that you do not know
whether you should begin by moving your pencil up — or down.[18] Again,
the function $f(x, y) = x^2 y / (x^4 + y^2)$ — with $f(0,0) = 0$ — has a nice,
pathwise connected, continuous looking graph, but it is not continuous in
Cauchy's sense.

Probably, in the future, definitions of continuity will be given that are
even more rigorous than those we have at present. Maybe they will be
more in line with our intuitions than the present definition is — but not
necessarily.

---

[18] An even trickier example is the 'Devil's staircase' function. For $x \leq 0$, let $f(x) = 0$.
For $x \geq 1$, let $f(x) = 1$. Otherwise, express x as a base 3 decimal, with no repeating 2
at the end. If there is no 1 in this base 3 number, convert all the 2's to 1's and interpret
the result as a base 2 number to get $f(x)$. If there is a 1, drop all digits after the first
1, and change any remaining 2's to 1's; then interpret the result as base 2 number to
get $f(x)$. For example, $f(1/2) = 1/2$ and $f(2/9) = 1/4$. This function is continuous, in
Cauchy's sense, at every point x. Its derivative is 0 everywhere it has a derivative, but
it raises, steadily, from 0 to 1, as x increases from 0 to 1. It cannot be drawn without
taking the pencil off the page.

6.3 Elementary–Advanced

If we use complex analysis to solve the Diophantine equation

$$x^2 + y^2 = z^2$$

we are applying an inordinately advanced method, since complex analysis presupposes a plethora of mathematical ideas, incorporating, and going far beyond, the few ideas necessary for solving this equation. However, if we use complex analysis to prove Cauchy's Residue Theorem, this is quite normal, since the theorem is, precisely, a theorem of complex analysis.

Other things being equal, it is better to have an elementary proof. Elementary proofs are more certain, more natural, and more accessible. They make more efficient use of our concepts, and sometimes lead to simplifications in the general theory related to what they prove. Thus even when they are longer, elementary solutions have a special merit.

On the other hand, a more advanced solution often gives an insight which an elementary solution cannot. With more advanced mathematics, sometimes, one can better grasp the underlying structure of the problem. Hence a solution can be censured as too elementary just as, in other cases, it can be condemned as unnecessarily advanced.

In some areas of mathematics, there is an ongoing rivalry between elementary and advanced solutions. This can be illustrated by some examples from the theory of numbers.

(1) Lagrange used simple continued fractions — and hence some analysis — to give the first complete solution of the Diophantine equation

$$x^2 - Ry^2 = C$$

(where R and C are any given integers). Thirty years later, in 1801, Gauss gave a more elementary solution which did not use any analysis. However, it is Lagrange's method which is the standard textbook method used today.[19]

(2) Van Lint conjectured that 120 is the only positive integer x such that $x + 1$, $3x + 1$, and $8x + 1$ are all squares. In 1969, Baker and Davenport used very advanced 'transcendental' number theory to prove this. In 1975, however, Kanagasabapathy and Ponnudurai gave a short, simple, and very elementary proof. Baker and Davenport had ascended a small hill using a

---

[19]C. F. Gauss, *Disquisitiones Arithmeticae,* trans. A. A. Clarke (New Haven: Yale University Press, 1966), sec. V.

high-altitude fighter plane.[20]

(3) Watson and others employed advanced mathematics to show that the only nontrivial positive integer solution of

$$x(x+1)(2x+1) = 6y^2$$

is $x = 24$ and $y = 70$. In 1990, Anglin, using the method of Kanagasabapathy and Ponnudurai, published a solution which is short and completely elementary.[21]

(4) In 1971, Finkelstein and London published a densely written page of algebraic number theory to show that

$$x^3 + 5 = 117y^3$$

has no solution in integers. However, this result follows in a much more elementary fashion from congruence considerations modulo 9. No integer of the form $x^3 + 5$ is evenly divisible by 9, but $117y^3$ is evenly disivible by 9. Hence there is no integer solution — as a college student might have told you for Monday's homework.[22]

(5) In 1988, Mohanty published a relatively elementary solution of the Diophantine equation

$$y^2 = x^3 - 4x + 1$$

A year later, Tzanakis and de Weger pointed out a fatal flaw in Mohanty's argument, and used the method of Baker and Davenport to give the first correct solution.[23]

---

[20] P. Kanagasabapathy, and T. Ponnudurai, 'The Simultaneous Diophantine Equations $y^2 - 3x^2 = -2$ and $z^2 - 8x^2 = -7$', *Quarterly Journal of Mathematics*, ser. 2, 26 (1975), 275-8.

[21] W. S. Anglin, 'The Square Pyramid Puzzle', *American Mathematical Monthly*, 97 (1990), 120-4.

[22] R. Finkelstein, and H. London, 'On D. J. Lewis's Equation $x^3 + 117y^3 = 5$', *Canadian Mathematical Bulletin*, 14 (1971), 111.

[23] N. Tzanakis, and B. M. M. de Weger, 'On the Practical Solution of the Thue Equation', *Journal of Number Theory*, 31 (1989), 99-132. Mohanty incorrectly assumed that if a square (e.g. 36) divides the product of two consecutive integers (e.g. 8 and 9) then it divides one of them.

(6) In 1988, Elkies showed how one could use the relatively advanced theory of elliptic curves to find a counterexample to Euler's claim that there are no positive integer solutions to

$$A^4 + B^4 + C^4 = D^4$$

Elkies had to admit, however, that Roger Frye had found a smaller (and hence more elegant) counterexample — using only the most elementary number theory.[24]

(7) In 1994, Wiles gave a proof of Fermat's Last Theorem using very advanced mathematics. For 350 years people had tried to find an elementary solution, and failed.

Elementary and advanced methods both have their merits. In one way it is interesting to watch David slay Goliath with a slingshot; in another way, it is interesting to watch David slay Goliath with a machine-gun. However, if the elementary solution is both shorter and more illuminating than the advanced solution, there is little question that the advanced solution is inferior. A proof that uses general topology to prove that there are infinitely many primes is unnecessarily advanced.

6.4 Simple–Complex

Other things being equal, human beings prefer the simple to the complex, and tend to assume that reality is as simple as is compatible with the observed facts. However, what, precisely, does it mean to say that a piece of mathematics is simple, or complicated? There are several answers to this question.

(1) A piece of mathematics is complex when clever mathematicians assert that it is hard to see what is going on in it. This is a subjective criterion.

(2) A piece of mathematics is complex if it is counter-intuitive, requiring the mathematician to examine it in a purely logical fashion.

---

[24] N. D. Elkies, 'On $A^4 + B^4 + C^4 = D^4$, *Mathematics of Computation*, 51 (1988), 825-35. In Frye's counterexample, $A = 95800$, $B = 217519$, $C = 414560$. Using only some 'congruence and divisibility properties', Frye showed that this is the smallest counterexample.

(3) A piece of mathematics is complex if there can be no short computer program which would cause it to be printed out, step by step, as a fully explicit, completely rigorous deduction in the language of set theory and logic. This is an algorithmic information criterion.[25] Note that a long proof could be produced by a short program if the long proof contained a lot of repetition or routine calculation. Note also that a short proof in, say, complex analysis, would require a long program for its production, because the concepts used in complex analysis rest on a mountain of definitions, each of which would have to be expressed in the language of set theory and logic.

(4) A piece of mathematics is complex if, relative to the level of mathematics at which it is found, it is long and involved. This definition of complexity allows us to say that complicated branches of mathematics contain simple proofs. In computing complexity in this sense, we do not add in the lengths of all the textbooks one would have to master to understand the concepts used. Indeed, it often happens that a concept at a more complicated level will simplify solutions to problems. For example, imaginary numbers make it easier to find real number solutions to cubic equations.

However we define it, complication is sometimes deadly. Consider Ellison's solution of the Diophantine equation $x^2 + 28 = y^3$. This solution is an ungainly combination of algebraic number theory, transcendental number theory, approximation theory, and computer calculation — and what began as a summer garden puzzle ends in a vast, haunted maze.

Simplicity can be deadly too. A proof which is too simple is boring. Francis Hutcheson writes:

> there are vastly greater pleasures in those complex ideas of objects which obtain the names of beautiful, regular, harmonious. Thus everyone acknowledges he is more delighted with a fine face, a just picture, than with the view of any one colour, were it as strong and lively as possible.[26]

The one colour would be monotonously simple. There would be no variety to combat the dull uniformity.

The twentieth century witnessed an interesting debate over complexity between two number theorists, L. J. Mordell and Serge Lang. In 1964, in

---

[25] G. Chaitin, 'Information-Theoretic Computational Complexity', in *New Directions in the Philosophy of Mathematics*, pp. 289-99.

[26] F. Hutcheson, 'An Inquiry Concerning Beauty', in *Aesthetic Theories*, ed. K. Aschenbrenner, and A. Isenberg (Englewood Cliffs: Prentice-Hall, 1965), p. 83.

reviewing Lang's *Diophantine Geometry,* Mordell rebuked Lang for employing a 'method of infinite assent in expounding his proofs, that is, simple ideas are often developed by using more complicated ones'.[27] Another number theorist, C. L. Siegel, agreed with Mordell, and wrote Mordell a letter in which he went so far as to describe Lang a 'pig'.[28] Lang, on his side, admitted that one of his goals was to 'jazz things up as much as possible' but he claimed there was nothing wrong about doing this in a research level monograph.[29] Since Mordell himself sometimes developed simple ideas 'using more complicated ones', it was really a matter of *how much* complexity one wanted to tolerate.[30] In any case, history decided in favour of Lang — in the sense that it turned out that the complicated methods he (and others) introduced into number theory led to many important results which were beyond the reach of Mordell's techniques. Wiles's proof of Fermat's Last Theorem was just one of several spectacular examples.

Perhaps from some aesthetic point of view, this is sad, and a mathematician who places a high value on simplicity might do well to avoid the latest research in the theory of numbers. On the other hand, there is no particular reason to think that the truth is always simple, and those who want the truth must be willing to accept it even when it is complicated.

6.5 Clarifying–Mystifying

Some proofs, even though they are written in a clear, detailed fashion, do not make us wiser. We see that the result is true, but we do not see why. Gauss's third proof of the Law of Quadratic Reciprocity falls in this category: it is mystifying.[31] After we study what Gauss wrote, we still wonder how reciprocity works. We cannot imagine how Gauss discovered his proof. Much research in number theory has been done precisely to end this puzzlement. Proofs of reciprocity have been found which make the phenomenon seem natural, inevitable, and almost commonplace. These

---

[27] L. J. Mordell, rev. of *Diophantine Geometry,* by S. Lang, *Bulletin of the American Mathematical Society,* 70 (1964), 495.

[28] S. Lang, 'Mordell's Review', *Notices of the AMS,* 42 (1995), 339.

[29] S. Lang, *Fundamentals of Diophantine Geometry,* 2nd edn. (New York: Springer-Verlag, 1983), p. 358.

[30] In order to deal with the rather simple idea of finding a square and a cube which differ by a given integer, Mordell used imaginary field ideals, and the ideal class group. See L. J. Mordell, 'The Diophantine Equation $y^2 - k = x^3$', *Proceedings of the London Mathematical Society,* 13 (1913), 60-80.

[31] Gauss discovered his third proof in 1808. It is the one usually given in elementary number theory textbooks.

proofs are at more advanced level, but they clarify reciprocity.[32]

Indeed, it is often the more advanced solution that does the clarifying. Elementary solutions tend to rely on tricks, whereas advanced solutions bring more general, more perspicacious concepts to bear on the problem. They situate it in a well-understood context, and relate it to powerful, well-understood ideas. For example, there is a solution to the Diophantine equation $x^2 + 2 = y^3$ which is very elementary, and may even have been known by Fermat. In effect, it relies on the fact that the class number of $\mathbf{Q}(\sqrt{2})$ is 1. Since, however, it makes no mention of this fact, it appears mysterious. There is something arbitrary or tricky about the elementary solution.

A proof can be mystifying in several ways. It can be expressed in turbid prose; it can be cluttered with inessential ideas; it can rely on advanced concepts which have not yet been made rigorous; or it can be too elementary to give the whole picture.

Some people prefer mystification. They enjoy the sensation of the numerological wonder that sometimes accompanies mathematical confusion. They desire, not plain, cold awareness, but warm, numinous bewilderment. Most mathematicians want clarity in mathematics, but one should note that this preference involves a value judgement related to the purpose or nature of mathematics. It is not a mathematical fact that clarity is better than mystification.

6.6 Specific–General

The best way to handle the specific problem of showing that

$$1 + 2 + 3 + \cdots + 10^{10} = 10^{10}(10^{10} + 1)/2$$

is by giving a simple proof of the more general fact that

$$1 + 2 + 3 + \cdots + n = n(n + 1)/2$$

On the other hand, there are instances where, although the general solution is difficult to follow, if one works the problem through for some special case then one immediately sees how to generalise to all cases. Hilbert's proof of the Dirichlet Unit Theorem furnishes an example. Due to unwieldy notation, it is bewildering. However, if one works it out for the case when the degree of the field is 3, it is then a perfectly routine matter to generalise the result to fields of degree n.

---

[32] K. Ireland, and M. Rosen, *A Classical Introduction to Modern Number Theory* (New York: Springer-Verlag, 1982), pp. 53, 73, 102, 108, 199, and 202.

Twentieth century mathematics is at once very general and very specific. At one extreme we have topology. At the other, we have a computer assisted proof that 112 is the smallest integer n such that a square of side n can be dissected into smaller squares, each with an integer side, and each different in size from all the others.[33] There is, moreover, a fruitful interaction between the general and the specific. For example, Alan Baker's general theorems in transcendental number theory can be applied to solve specific Diophantine equations.[34]

In 1900, at the International Congress in Paris, David Hilbert posed twenty-three problems, the tenth of which is to find a general method for solving any given Diophantine equation. In 1970, Yuri Matijasevic proved that there is no such general method, but research is none the less conducted in the hope of approximating Hilbert's ideal.[35] S. Chowla, in reviewing a paper in which Baker published a solution of a large class of Diophantine equations, exclaims that Baker 'has solved Hilbert's tenth problem in a special case'.[36]

One can pay a price for general solutions. They are sometimes so general that it is difficult to apply them in specific cases. Brindza calls his article 'On Some Generalisations of the Diophantine Equation

$$1^k + 2^k + \cdots + x^k = y^z \text{ '}$$

However, the generalisations in question are useless for solving the equation when k and z are both 2. For that, the reader is referred to earlier papers, which contain wholly different methods.[37]

6.7 Concrete–Abstract

To abstract from something is to ignore its details and concentrate on its underlying structure. And the decision about what constitutes a detail and what belongs to the underlying structure may involve a value judgement.

---

[33] This theorem was first proved by A. J. W. Duijvestijn. The 21 smaller squares have sides 2, 4, 6, 7, 8, 9, 11, 15, 16, 17, 18, 19, 24, 25, 27, 29, 33, 35, 37, 42, and 50. Now that's a jigsaw puzzle!

[34] A. Baker, and H. Davenport, 'The Equations $3x^2 - 2 = y^2$ and $8x^2 - 7 = z^2$', *Quaterly Journal of Mathematics,* ser. 2, 20 (1969), 129-37.

[35] M. Davis, 'Hilbert's Tenth Problem is Unsolvable', *American Mathematical Monthly,* 80 (1973), 233-69.

[36] S. Chowla, rev. of 'Contributions to the Theory of Diophantine Equations', by A. Baker, *Mathematical Reviews,* 37 (1969), 737.

[37] B. Brindza, 'On Some Generalisations of the Diophantine Equation $1^k + 2^k + \cdots + x^k = y^z$', *Acta Arithmetica,* 44 (1984), 99-107. Brindza incorrectly ascribes a solution of the special case to E. Lucas.

For example, two dimensional vector arithmetic is an abstraction from the addition of net physical displacements. The vector space is an abstraction from two dimensional vector arithmetic. The module is an abstraction from the vector space. What is, at one level, part of the underlying structure is, at another level, just a further detail to be removed.

Often the more abstract structure is instantiated in numerous 'concrete' situations, and hence more abstract usually implies more general. The converse is false. Euler's equation

$$a^{\phi(m)} \equiv 1 \ (\mathrm{mod} \ m)$$

is more general than Fermat's 'Little Theorem'

$$a^{p-1} \equiv 1 \ (\mathrm{mod} \ p), \quad \text{with } p \text{ prime}$$

but it is at the same level of abstraction. Many of the informal proofs in calculus are quite general, although they are also quite concrete. Archimedes's 'centre of gravity proof' that the three medians of any triangle are concurrent applies to Euclidean triangles generally, but it is none the less concrete.[38]

Which is better, abstraction or concreteness? Do we want category theory, or merely a better algorithm to factor large integers? According to some religions, it is a happy thing to escape the concrete world. Studying abstract mathematics might be viewed as an exercise highly beneficial for aspiring mystics. However, those who venerate the details of the tangible world may feel that abstract mathematics is justified only insofar as it may have some application to problems in the physical sciences.

Davis and Hersh, in *The Mathematical Experience,* claim that the computer is the cause of

> greater interest in constructive and algorithmic results, and decreasing interest in purely existential or dialectical results that have little or no computational meaning.[39]

With regard to the 'current trend of extreme abstraction', they write:

> Researchers carrying out an ultra-abstract program frequently devote the bulk of their effort to straightening out difficulties in the terminology they have had to introduce, and the remainder of their effort to reestablishing in camouflaged form what

---

[38] T. Heath, *A History of Greek Mathematics,* vol. 2 (New York: Dover, 1981), pp. 75-78.

[39] P. J. Davis, and R. Hersh, *The Mathematical Experience* (Boston: Birkhäuser, 1981), p. 16.

has already been established more brilliantly, if more modestly. Programs of extreme abstraction are frequently accompanied by attitudes of complete hauteur on the part of their promulgators, and can be rejected on emotional grounds as being cold and aloof.[40]

Along the same lines, Morris Kline says:

Generalization and abstraction undertaken solely because research papers characterized by them can be written are usually worthless for application. In fact, most of those papers are devoted to a reformulation in more general or more abstract terms or in new terminology of what had previously existed in more concrete and specific language. And this reformulation provides no gain in power or insight.[41]

E. T. Bell takes a more moderate view. He agrees that abstraction can prove barren, but he also thinks that

The abstract, postulational method is not mere cataloguing and pigeon-holing. It also is creation, but of a kind more basic than the disorderly luxeriance of the nineteenth century.[42]

It should also be noted that some very abstract theory was used to prove Fermat's Last Theorem.

In evaluating a work of mathematics in terms of the concrete–abstract dichotomy, we cannot simply say that 'abstract is good' or 'abstract is bad'. We need to think about the level of abstraction in relation to the type of mathematics under consideration, and in relation to the point or spirit of what is being done. It would be inappropriately abstract to use Lagrange's Theorem from group theory to prove a result in number theory which could be established in a much more elementary fashion, but it might also be inappropriately concrete to use Lagrange's Theorem to demonstrate a result in category theory.

6.8 Constructive–Nonconstructive

'Constructive' has several meanings. It can refer to a finite sequence of straightedge and compass operations in Euclidean geometry. It can refer

---

[40] Ibid., p. 116.

[41] M. Kline, *Mathematics: The Loss of Certainty* (Oxford: Oxford University Press, 1980), p. 283.

[42] E. T. Bell, *The Development of Mathematics,* 2nd edn. (New York: McGraw-Hill, 1945), pp. 261 and 267.

to existence proofs which enable one to calculate the objects in question. It can refer to logical procedures which do not depend on the Law of the Excluded Middle. These different meanings of 'constructive' have something in common: nothing is 'constructive' unless it is straightforwardly accessible to finite minds.

Archimedes's derivation of the formula for the area a circle is nonconstructive in the sense that he fails to give a Euclidean construction for his segment of length π, and also in the sense that he relies on the Law of the Excluded Middle.[43] Cantor's proof of the existence of nonalgebraic reals is nonconstructive in the sense that, although it shows that there are 'more' reals than algebraic numbers, it does not give us a way of pointing to even a single example of a nonalgebraic real. Thue's proof that there is a bound M on the integer solutions of

$$ax^3 + bx^2y + cxy^2 + dy^3 = e \qquad \text{with } ade \neq 0$$

is nonconstructive in the sense that it does not yield a finite means for calculating this bound. Van Dalen's proof that there are irrationals a and b such that a^b is rational is nonconstructive in the sense that it gives us no effective procedure for finding examples of such irrationals.[44]

Some mathematicians believe that mathematical objects exist independently of human beings. From this point of view, it makes sense to talk about the existence of a mathematical object, even if there is no finite way of exhibiting it. Constructivists reject this. For a constructivist, 'to be is to be constructible'. Some constructivists go so far as to claim that a mathematical object begins to exist only on the day a human mathematician first discovers how to construct it. Bishop and Bridges claim that the set A of twin primes is not infinite because we do not know that it is infinite. However, 'this does not rule out the possibility that at some time in the future A will have become countably infinite'.[45] Most mathematicians find this an odd way of speaking.

In any case, there are advantages to having a constructive proof. It is good to be able to 'see what one is talking about'. In 1968, Baker published a constructive proof of Thue's Theorem, giving a finite procedure for actually calculating the bound M (in terms of the given coefficients). Baker's bound was astoundingly large, and his proof was more complicated

---

[43] Archimedes, *The Works of Archimedes,* ed. T. L. Heath (New York: Dover, 1897), pp. 91-8.

[44] Van Dalen noted that if $c = \sqrt{2}^{\sqrt{2}}$ then one of c and $c^{\sqrt{2}}$ is the required rational — but we are not told which!

[45] E. Bishop, and D. Bridges, *Constructive Analysis* (Berlin: Springer-Verlag, 1985), p. 18.

than that of Thue. None the less, mathematicians hailed Baker's work as an important improvement.

6.9 Practical–Impractical

A solution can be practical in the sense that it gives a practical way of actually computing the mathematical object in question (for example, the formula which represents a given curve). Or it may be impractical in that, say, the known universe is not large enough to contain a computer with enough memory space to manipulate the numbers required in the calculation. The factorisation of large integers is a popular topic at the moment, and practical algorithms exist for factoring integers with 100 digits.[46] When the numbers have 50,000 digits, however, the software runs too slowly to be of any interest. A paradigm of impracticality is Baker's solution of the Diophantine equation $x^2 + k = y^3$ (where k is a given nonzero integer). Baker solves this equation by proving that the absolute values of the integers x and y must be less than

$$e^{(10^{10}|k|)^{10000}}$$

where $e = 2.72\ldots$[47]

What if there were no distinction between practical and impractical in mathematics? For example, suppose we could perform a countably infinite number of mathematical operations in a finite amount of time. (We might perform our calculations at an accelerating speed: one second for the first addition, half a second for the second addition, a quarter of a second for the third addition, and so on.) With such an ability, a bound like Baker's would be no more formidable than a bound of fifty. Indeed, in just two seconds we could check all quadruples (x, y, z, n) of positive integers to show that we can never have

$$x^{n+2} + y^{n+2} = z^{n+2}$$

— thereby giving a quick, elementary proof of Fermat's Last Theorem.

Moreover, consider any conjecture T. If T has a proof then that proof can be written down using the 100 symbols on a typical keyboard (including the space). The proof can thus be identified with a natural number written in base (or scale) 100. Hence to find a proof of T, one need only look at each of the natural numbers, written in base 100, and read it. This involves only

---

[46] In 1994 Arjen Lenstra and a group of about 1700 Internet users succeeded in factoring a 129 digit number composed of two large primes. This number, called RSA129, had been given as a factoring challenge by R. Rivest, A. Shamir, and L. Adelman. They claimed it would take 40 quadrillon years to factor their number, but, in fact, it only took 17.

[47] Baker and Davenport, 'The Equations $3x^2 - 2 = y^2$ and $8x^2 - 7 = z^2$'.

a countably infinite number of operations. Thus, if there were no distinction between practical and impractical in mathematics, it would take only two seconds to find a proof for any given conjecture, or else establish that there is no proof.

Nor would this way of doing mathematics make it trivial. For, relative to any typical set of axioms, there are statements which can be neither proved nor disproved. Given such a statement, we would want to think about choosing a new axiom from which it, or its negation, would follow. Since there is no mechanical procedure for judging an axiom as a 'good' one, the choice would require artistic or philosophic input. Of course, a mathematician who could do a countably infinite number of things in a finite time could review all possible axiom systems. Finding enlargements to the original system would not be the problem. Rather the problem would be to find an enlargement that was elegant or simple or intuitive or true — and answered important questions from the branch of mathematics under consideration. Without the distinction between practical and impractical, mathematicians would be more like philosophers.

Of course, this is fantasy. Given the current human condition, the practical–impractical distinction in mathematics is far from dispensable, and our axiology ought to acknowledge the benefits of practicality. A Euclidean construction which is so lengthy that it is practically impossible to use it to produce an accurate drawing of the figure is not as good as a short construction for the same diagram. A method of factoring integers which takes two years to factor a number is not as good as one which factors the same number in two minutes.

6.10 Elegant–Ugly

As the tenth axiological dichotomy, we have, not beautiful–ugly, but elegant–ugly. This presupposes that elegance is the only sort of beauty in mathematics. H. E. Huntley and F. Le Lionnais would dispute this.

In *The Divine Proportion*, Huntley asserts that a piece of mathematics may be beautiful, not because it is elegant, but because it puts us in touch with something deep within ourselves. He claims that geometry is sometimes beautiful, not because it is elegant, but because the curves are 'sensuous'.[48]

This is wrong. If a sensuous curve is beautiful, this is not because it reminds us of, say, a breast, but because it has a symmetry or grace which instantiates elegance. Huntley is also wrong to connect beauty to what puts us in touch with something deep within ourselves. Consider, for example,

---

[48] H. E. Huntley, *The Divine Proportion* (New York: Dover, 1970), pp. 17 and 86.

some tedious problem about the trajectory of cannon-balls. Such a problem might well put us in touch with a deep, inner violence, but none the less be totally lacking in beauty.

Le Lionnais distinguishes 'romantic beauty' from elegance. Mathematics has romantic beauty insofar as it glorifies 'violent emotion, nonconformism and eccentricity'.[49] Le Lionnais cites continuous nondifferentiable functions as an example of this sort of beauty. For Le Lionnais, fractals, such as the von Koch snowflake, are not elegant, but romantically beautiful.

This also is wrong. Fractals, no less than other geometric objects, are beautiful just insofar as they are elegant. The von Koch snowflake is beautiful because it is an elegant balance of complexity and simplicity: its intricate boundary gives it variety, while its symmetry and self-similarity endow it with uniformity. It excites without overwhelming.

Now it is true that certain proponents of fractals have a 'romantic' taste in colour. They like to look at orange von Koch snowflakes on purple backgrounds. This is certainly eccentric, and one might well talk about the 'violent' beauty of the colour schemes. The fractals themselves, however, cannot be blamed for the garish manner in which they are coloured. As invisible mathematical objects, they are beautiful simply because they are elegant.

We conclude, then, that in mathematics the opposite of the ugly is, indeed, the elegant.

What exactly does one mean by elegance? Consider the problem of adding up the integers from 1 to 100. One way to do this is simply to add them up, using the algorithm Miss Grundy teaches in kindergarten. Another way to do it is as follows:

$$
\begin{array}{ccccccccccc}
1 & + & 2 & + & 3 & + & \cdots & + & 49 & + & 50 \\
100 & + & 99 & + & 98 & + & \cdots & + & 52 & + & 51 \\
\hline
101 & + & 101 & + & 101 & + & \cdots & + & 101 & + & 101 & = & 50 \times 101
\end{array}
$$

Now, although there is often a subjective component in aesthetic evaluation, every mathematician, the world over, will tell you that the second solution is more elegant than that of Miss Grundy. The question is not 'which solution is more elegant?' but 'how is elegance identified?' or 'what are the defining marks of elegance?'

One answer to this question is the following. A piece of mathematics is elegant if and only if it is

---

[49] F. Le Lionnais, 'Beauty in Mathematics', in *Great Currents of Mathematical Thought,* vol. 2, ed. F. Le Lionnais, trans. C. Pinter and H. Kline (New York: Dover, 1917), p. 130.

well-expressed
brief
efficient
illuminating
simple
balanced
unifying, and
exciting

Let us consider these eight marks of mathematical beauty in turn.

Well-expressed

Against the view that mathematics is beautiful only if it is well-expressed, one might object that this confuses the elegance of the mathematics with the elegance of the prose expressing it. To this I reply that if one has a clear mathematical insight, uncluttered by extraneous material, one will ordinarily express it in a limpid, well-organised, and hence attractive prose style. Beautiful ideas engender felicitous writing. Besides, what we are evaluating when we judge a piece of mathematics is not the beauty of whatever it is the mathematician senses in his or her heart, but the beauty of what is actually offered to us in the lecture or article. A bad painting of a beautiful model is still a bad painting.

Brief

In *De Poetica* 1450b–51a, Aristotle writes:

> to be beautiful, a living creature, and every whole made up of parts, must not only present a certain order in its arrangement of parts, but also be of a certain definite magnitude. Beauty is a matter of size and order, and therefore impossible ...in a creature of vast size — one, say, 1000 miles long — as in that case, instead of the object being seen all at once, the unity and wholeness of it is lost to the beholder.

In a similar vein, Kant opposes beautiful magnitudes to sublime magnitudes:

> for the mathematical estimation of magnitude there is, indeed, no maximum (for the power of numbers extends to infinity); but for its aesthetical estimation there is always a maximum, and of this I say that, if it is judged as the absolute measure

than which no greater is possible subjectively (for the judging subject), it brings with it the idea of the sublime.[50]

Although Miss Grundy's solution is not as long as many an unwieldy monster found in contemporary mathematics journals, it is still too long for what it does. In the second, short solution, we see *why* the answer is 5050, but in the solution by brute calculation, we are overwhelmed by detail. 'The wholeness of it is lost to the beholder'.

To this it may be objected that size is relative. A large object can be beautiful if seen from the right perspective. An astronaut can look at a whole planet, and find it beautiful.

The answer to this objection is that beauty, too, is relative. When humans call something elegant, we do not mean that it ought to seem so to an angel. We are perfectly within our conceptual rights to insist that a mathematical solution which is too big for *us* is, for that reason, inelegant. Coxeter writes, 'a proof is "elegant" for me if it involves fewer steps than *I* would have expected.'[51]

Efficient

A proof is efficient if a powerful or deep theorem is proved with relatively little effort, or with a minimal amount of advanced mathematics. For example, to prove Fermat's conjecture that every natural number is a sum of three triangular numbers, it suffices to show that every natural number of the form $8M + 3$ is a sum of three squares. It is not necessary to get tangled up in the longer, more complex 'three squares theorem' for natural numbers in general. Relative to the goal of proving Fermat's conjecture, it would be inefficient to go through a long demonstration of the fact that numbers of the form, say, $8M + 2$ can be written as sums of three squares.

Illuminating

To be elegant, a proof needs a high ratio of understanding to length. Miss Grundy's solution is ugly partly because it fails to explain *why* the answer is 5050. Nor does it offer any insight into how to add up longer sequences of consecutive integers where the Grundy method would be hopelessly impractical. One of the reasons the short solution is elegant is that gives us an *insight* into arithmetic progressions generally.

---

[50] I. Kant, *Critique of Judgement,* trans. J. H. Bernard (New York: Hafner, 1951), pp. 89-90 (section 26).

[51] H. M. S. Coxeter, in a private letter to Anglin, dated October 29, 1989.

Simple

Excessive simplicity is a blemish. A proof needs a surprise construction or a subtle twist to rouse our aesthetic admiration. Excessive complexity, on the other hand, is fatal. The golden mean is not a moderate degree of complexity so much as a moderate degree of simplicity.

Balanced

A proof is not well balanced if it contains five cases, four of which involve brief remarks from Galois theory, and one of which is a long computation in analytic geometry. In an elegant piece of mathematics, the structure of the lemmas, theorems, and corollaries is such that they correspond to each other in some pleasing proportion. Calculations are balanced by insights, and abstract generalisation by meaningful applications.

Unifying

A piece of mathematics is unifying if it organises the various givens into some harmonious whole, if it links different cases, locating the same structure in different phenomena, or if it reveals links between two very different branches of mathematics.

However it is accomplished, the uniting of disparate elements has aesthetic appeal. To take one example, the Dandelin sphere derivation of the formula for the ellipse is elegant partly because it evokes the symmetries of the sphere, and partly because it relates the ellipse problem to the equivalent problem for the hyperbola, providing the key to both problems at once.

Exciting

No one thinks a boring proof is elegant, but a proof which builds to a climax, whose dénouement is achieved by a series of deft surprises, whose conclusion conjures up a bevy of charming corollaries — such a proof is a gem.

Again, the Dandelin sphere treatment of the ellipse provides an example. Why are those spheres there? How will they be used? Suddenly, a pivotal equation appears! The required answer falls out. Then the spheres locate the foci. And the same spheres steal the secret of the hyperbola.

6.11 Meaningful–Pointless

A solution in mathematics is pointless insofar as it is an isolated curiosity, meaningful insofar as it expresses deep ideas — whether mathematical, scientific, or philosophical. In particular, a piece of mathematics is meaningful if

(1) it places the problem in an illuminating context, or
(2) it is natural as opposed to artificial, or
(3) it symbolises some important aspect of human experience, or
(4) it is historically significant, or
(5) it opens us to the infinite.

Illuminating

In *A Mathematician's Apology*, G. H. Hardy writes:

> a mathematical idea is 'significant' if it can be connected, in a natural and illuminating way, with a large complex of other mathematical ideas.[52]

Suppose the problem is to show that

$$1/2 + 1/3 + 1/4 + \cdots + 1/20 < 3$$

A pointless way of doing this is simply to add up the fractions. A meaningful way to solve the problem is to put it in the context of area and integration:

$$1/2 + 1/3 + 1/4 + \cdots + 1/20 < \int_1^{20} \frac{1}{x}\, dx = \log_e 20 = 2.9957 < 3$$

(The area bounded by the lines $x = n$, $x = n+1$, the x-axis, and the curve $y = 1/x$ contains a rectangle with base 1 and height $1/(n+1)$.)

Note that the illuminating context in which a meaningful solution places a problem need not be purely mathematical. It might be artistic or chemical. The theory of differential equations, for example, is meaningful partly because of its applications in science.

Natural

In 1971, Finkelstein and London published a proof of the fact that

$$x^3 + 117y^3 = 5$$

---

[52] G. H. Hardy, *A Mathematician's Apology* (Cambridge: Cambridge University Press, 1967), p. 89.

has no integer solutions.[53] Their proof consisted of a densely written page of algebraic number theory — a very complicated and artificial tool for a straightforward problem. The natural, and hence meaningful, proof can be written in two lines:

since $x^3 \equiv 5 \pmod 9$ has no solution,
$x^3 + 117y^3 = 5$ has no solution.

This shorter, more elementary proof expresses the important way in which divisibility governs integer solutions to polynomial equations.

Symbolic

The problem of Apollonius (225 BC) is the problem of finding a straightedge and compass construction for a circle tangent to three given circles. In terms of Carl Jung's system, the smallest of the three given circles symbolises the ego-consciousness, while the other two symbolise the shadow, and the anima or animus. The construction of the fourth circle symbolises the individuation process, with the fourth circle representing the Self. The geometric drawing, with the four circles, is a mandala.[54]

The key to a quick, elegant solution of the problem of Apollonius is the notion of inversion. In Jungian terms, the circle of inversion stands for the person. Inversions from the inside to the outside of that circle symbolise psychic projections. The inversion construction consists of solving the problem outside the circle of inversion, and then taking the solution back inside that circle. This can be interpreted as the person achieving individuation (psychic harmony) by means of engaging in some activity in the external world, such as art or psychoanalysis. On Jung's view, actually carrying out the geometric construction can help one attain the psychic harmony symbolised by the resulting diagram.[55]

Many theorems in Euclidean geometry can be psychoanalyzed. The equality of different angles on the same side of the same chord of a circle symbolises constancy in the midst of vacillation. The existence of certain invariants, such as the sum of the angles of any triangle, points to stability obtained by compensation. The nine-point circle suggests unity and order in the midst of apparent chaos. Indeed, since the nine-point circle provides a smooth, circular path from the incircle to any of the excircles, it symbolises a unity and order in a person that joins the inner and the outer 'man'. The

---

[53] Finkelstein and London, 'On D. J. Lewis's Equation $x^3 + 117y^3 = 5$'.

[54] C. G. Jung, 'Psychology of the Transference', in *The Collected Works of C. G. Jung*, vol. 16, trans. R. F. C. Hull (New York: Pantheon Books, 1960), pp. 207, 317, and 321.

[55] A good account of the inversion construction is found in C. S. Ogilvy, *Excursions in Geometry* (New York: Oxford University Press, 1969).

possibility of placing an icosahedron inside an octahedron, with one vertex on each edge of the octahedron, evokes the possibility of an outwardly simple person having a rich and complex inner life. The vertices of the icosahedron divide the edges of the octahedron in the golden mean: such a person has achieved balance and happiness.

Needless to say, the geometer may not be aware of these associations, but, precisely, they act on the level of the unconscious, giving the mathematics its attraction and fascination.

Historic

Another factor contributing to meaningfulness in mathematics is historical or cultural significance. For example, it would be meaningful to find a proof of the Erdös conjecture that, given any positive integer n (greater than 4), there are distinct positive integers x, y, and z such that

$$\frac{4}{n} = \frac{1}{x} + \frac{1}{y} + \frac{1}{z}$$

This would be meaningful because the subject of 'unit fractions' goes back to ancient Egypt.

Mathematicians are sometimes guilty of ignoring history and culture. If an art critic maintains that ancient or medieval art no longer has any value, we protest that this is narrow, but when the Mathematics Department establishes its course requirements in such a way that nothing is learned unless it is directly tied to current research, we believe that this is all right. A great deal of meaningful mathematics is, in fact, left to the few historians of the subject — as if the world's curators had decided that the only important paintings were twentieth century paintings.

Infinite

Finally, a piece of mathematics is meaningful if it removes our horizons, opening us to the infinite. A good example is Cantor's proof that there is an infinite hierarchy of infinite cardinals. Such a proof puts us in touch with the infinite within ourselves.

6.12 Fruitful–Sterile

There are at least five ways in which a piece of mathematics can be fruitful.

(1) A piece of mathematics is fruitful if it has applications in science or technology. For example, a new proof technique is fruitful if it can be used to increase the speed of an important computer package.

(2) A theorem is fruitful if, as an historical fact, it led mathematicians to some meaningful, new result. Note that if results R and R' are logically equivalent, it may be an historical accident that R' was the fruit of R rather than the other way round.

(3) A method or result is fruitful if, in a good, axiomatic or pedagogical presentation of the relevant topic, it naturally goes at the beginning, with key theorems being derived from it. In this sense the side-side-side congruence theorem is a fruitful theorem in Euclidean geometry.

(4) A piece of mathematics is fruitful if it drives mathematicians to do research. Zeno's paradoxes inspired centuries of work on the continuum. The Russell paradox provoked intense work in foundations.

(5) A bit of mathematics is fruitful if it makes results accessible to nonspecialists, inspiring them to consider working in the related branch of mathematics. Ogilvy's clear presentation of inversion stirs up a desire to do synthetic geometry. Martin Gardner's puzzles cause young minds to choose careers in mathematics.

Note that 'fruitful' does not mean 'conducive to publishing papers in my sub-specialty'. Some mathematical journals, as a matter of policy, will publish a turbid (but original!) proof of some nugatory generalisation rather than a brilliant exposition of a well-known, important result. The author then gets tenure to go on developing his useless ideas, while those who would have benefited from the brilliant exposition are left in ignorance. Rota deplores

> the simplistic view of mathematics as a linear progression of problems solved and theorems proved, in which any other function that may contribute to the well-being of the field (most significantly, that of exposition) is to be valued roughly on a par with that of a janitor.[56]

6.13 Conclusion

This concludes our examination of the twelve dichotomies. In the next chapter we shall explain how they can be coordinated to give an overall picture of the net worth of a piece of mathematics.

---

[56] G. Rota, rev. of *I Want to Be a Mathematician*, by Paul R. Halmos, *American Mathematical Monthly*, 94 (1987), 700-2.

Chapter 7

Mathematics and Values II

7.1 The Rating Scheme

To give an overall rating to a piece of mathematics, we look at each of the twelve dichotomies in turn. If we think that, on the whole, the mathematics does well in terms of a dichotomy — and we should have an explicit reason for thinking so — we give it a 1.[1] If, however, we think that, on the whole, the mathematics does badly when judged in terms of a dichotomy, we give it a −1. If, as often happens, we cannot give it a 1 or a −1, we give it a 0 for the dichotomy in question. Thus, if a piece of mathematics is in every respect perfect, it scores 12, but if it is in every respect dreadful, it scores −12.

As an example, we give a rating for the short solution, given above, to the problem of adding all the integers from 1 to 100.

---

[1] This does **not** mean that we think the mathematics exemplifies, say, the first member of the dichotomy, but that we think it is satisfactory from the point of view of the two opposites considered together.

SOLUTION: Short solution of $1 + 2 + \cdots + 100 = x$

Dichotomy	Points	Reasons
Complete–Incomplete	0	
Rigorous–Intuitive	1	it is both
Elementary–Advanced	0	
Simple–Complex	1	a slight twist
Clarifying–Mystifying	1	explains why sum is 5050
Specific–General	0	
Concrete–Abstract	0	
Constructive-Nonconstructive	0	
Practical–Impractical	1	faster than adding singly
Elegant–Ugly	1	a neat trick
Meaningful-Pointless	0	
Fruitful–Sterile	1	applies to any AP

When one does a rating, it is tempting to add up the points. Here one must be careful. Are the values represented by the points commensurable? Does a thing have overall goodness in direct proportion to the number of good qualities it has? Should a very elegant proof be given 2 points for elegance? Should the categories be weighted, since, say, completeness in a proof is three times more important than fruitfulness? I do not think it is impossible to justify point addition, but I have not found an easy and convincing way of doing so, and my own ratings culminate, not in single numbers, but in twelve-component vectors.

Note that my unwillingnes to add up the points does not imply that I am unwilling to say that one piece of mathematics is simply better than another. Suppose we have two proofs, $P1$ and $P2$, of the same theorem, and suppose that $P1$ beats $P2$ in at least one of the twelve categories, while $P2$ does not beat $P1$ in any of the twelve categories. Then I would be quite happy to conclude, not merely that $P1$ is better, in some respects, than $P2$, but also that it is better *simpliciter* than $P2$. There is, of course, the possibility that $P2$ beat $P1$ with respect to some thirteenth dichotomy, not on our list, but I am not inclined to think that this thirteenth dichotomy would be as important as the others.

7.2 The Example of the Medians

In this section we use the above axiology to evaluate and compare five proofs of the theorem that, in any triangle, the medians meet at a single point.[2]

(1) Archimedes's Proof

Think of a triangle ABC as a thin sheet of metal of uniform thickness and density. Think of it as made of thin, nonoverlapping strips of equal, uniform width, all parallel to BC. From the theory of similar triangles, it follows that the midpoints of these strips lie on the median through A. Since each strip has uniform thickness and density, its balance point is its midpoint. Thus, if we want to balance the triangle on the edge of a ruler, we can do so by placing the median through A, and hence all the midpoints of the strips, on the edge of the ruler. Indeed, this is the only way of balancing the triangle on the ruler's edge, if we want point A on the ruler's edge. For, since the triangle balances on the median, it will not balance on any other line through A. From this it follows that if we want to balance the triangle on the tip of a pencil, we must place some point of the median through A on the tip of the pencil. (For if the triangle balanced on a point X not on that median, then it would balance on the line AX, against what we just proved.) Hence the 'centre of gravity' of the triangle lies on the median through A. Similarly, it lies on the median through B, and on the median through C. Thus the centre of gravity of the triangle is a point through which the three medians pass.

Archimedes's proof is not mathematically rigorous. It makes unjustified assumptions about thin strips and continuity. On the other hand, it is intuitive in a way that is convincing and illuminating. Its very concreteness clarifies the situation. We can thus give it points in the categories 'rigorous–intuitive', 'clarifying–mystifying', and 'concrete–abstract'. Expressing a beautifully simple but practical procedure, Archimedes's proof also deserves points in the categories 'simple–complex' and 'practical–impractical'. For linking geometry and statics, Archimedes's proof also merits points for meaningfulness and fruitfulness. The rating for this proof is summarised in the first column of the table below.

(2) A Euclidean Proof

Let triangle ABC have medians AD, BE, and CF. Suppose the medians are not concurrent, but form a little triangle XYZ with X on AD and BE, and Y on BE and CF, and Z on AD and CF. Without loss of generality, we take it that X is in the interior of $\angle ACF$, so that Z is in XD and Y is

---

[2] A median is a straight line segment joining a vertex to the midpoint of the side of the triangle opposite that vertex.

in XB. Since they have equal bases and the same height, triangles BZD and ZDC have the same area r. Let r' be the area of $BYZD$. Then $r' > r$.

Similarly, if s is the area of YFB and YFA, while s' is the area of $AXYF$, we have $s' > s$. Also if t is the area of XEA and XEC, while t' is the area of $CZXE$, we have $t' > t$. Hence

$$r' + s' + t' > r + s + t$$

Since triangles ADB and ADC are equal in area,

$$r' + s + s' + XYZ = r + t + t'$$

Since triangles BEC and BEA are equal in area,

$$r' + r + t' + XYZ = s' + s + t$$

Adding these two equations, and simplifying, we obtain

$$t = r' + XYZ$$

so that $t > r'$. Similarly, $r > s'$ and $s > t'$. Thus

$$r + s + t > r' + s' + t'$$

Contradiction. Hence the medians are concurrent.

This proof is complete and rigorous. Requiring only the most basic ideas about areas, it is delightfully simple. It is mystifying, however, since it does not give any clue as to *why* the medians are concurrent. Nor does it tell us exactly where they intersect. We must do some further work to 'construct' the point of concurrency. On the other hand, the proof is brief and neat; the adding of the equations with the result

$$t = r' + XYZ$$

is a pleasant surprise; hence one can call the proof elegant. A rating is found in column 2 of the table below.

(3) Another Euclidean Proof

Suppose that the medians BE and CF of triangle ABC meet at G. Let H be the midpoint of BG, and K the midpoint of CG. HK is then half as long as BC and parallel to it. But FE is also half as long as BC and parallel to it. Hence $FHKE$ is a parallelogram. Since the diagonals of a parallelogram bisect each other, $GH = GE$. Thus, since H is the midpoint of BG, G trisects BE, with $BG > GE$. The median CF trisects the median

BE (with the trisection point further from B than from E). Similarly, the median AD trisects BE (with the trisection point further from B than from E). Thus all three medians pass through this trisection point.

The above proof is tricky, because it depends on the idea of constructing HK. No hint is given as to how one would hit on this idea: it would be a 'stroke of genius', or an 'inspiration'. The construction's cleverness makes the proof mystifying, but in a way that adds to, rather than undermines, its elegance.

The proof is constructive in the sense that it gives us a practical way of locating the point of concurrency — two thirds of the way down a median starting from its vertex. This extra information was not available in the previous two proofs, and hence this proof is more complete and illuminating.

(4) A Vector Proof

Suppose the triangle's vertices are the tips of the vectors $\vec{a}$, $\vec{b}$, and $\vec{c}$. Then the midpoints of the three sides of the triangle are the tips of vectors $(\vec{a} + \vec{b})/2$, $(\vec{a} + \vec{c})/2$, and $(\vec{b} + \vec{c})/2$. Since

$$\begin{aligned}
(\vec{a} + \vec{b} + \vec{c})/3 &= (2/3)(\vec{a} + \vec{b})/2 + \vec{c}/3 \\
&= (2/3)(\vec{a} + \vec{c})/2 + \vec{b}/3 \\
&= (2/3)(\vec{b} + \vec{c})/2 + \vec{a}/3
\end{aligned}$$

it follows that the point at the tip of $(\vec{a} + \vec{b} + \vec{c})/3$ is $2/3$ the way along each median, going from the vertex to the opposite midpoint. Hence the point at the tip of $(\vec{a} + \vec{b} + \vec{c})/3$ is on all three medians.

This proof is mystifying in that it does not tell us where the idea of using the vector $(\vec{a} + \vec{b} + \vec{c})/3$ comes from. On the other hand, it does tell us exactly where the point of concurrency is. The proof requires a knowledge of vectors, and is, perhaps, inappropriately advanced for the very elementary result, but, in its favour, it makes very good use of its vectors, and exemplifies a technique that can be used to prove other problems in Euclidean geometry. A rating is found in the fourth column of the table below.

(5) An Analytic Geometry Proof

Without loss of generality, let the triangle have vertices $(0, 0)$, $(2, 0)$, and (s, t). Then the medians are

$$\text{(a)} \quad y = \frac{t}{s + 2} x$$

$$\text{(b)} \quad y = \frac{t}{s-1} x - \frac{t}{s-1}$$

$$\text{(c)} \quad y = \frac{t}{s-4} x - \frac{2t}{s-4}$$

(a) and (b) intersect in a point P with x-coordinate $(s+2)/3$. (a) and (c) intersect in a point Q with x-coordinate $(s+2)/3$. Since the x-coordinates are the same, and both P and Q are on (a), it follows that P and Q are the same point. The three medians thus meet at this point.

By the time one has checked all the algebraic details, one realises that this proof is too long. Also, it is factitious. It does not explain anything, but merely grinds out the result, using excessively heavy machinery. The algebraic symbolism hides, rather than reveals, the location of the point of concurrency. If one wants an example of a vapid exercise, devoid of ingenuity, this is it.

We now give a table summarising our evaluations.

SOLUTION: Five Proofs of Median Concurrency

Dichotomy			Points		
	1	2	3	4	5
Complete–Incomplete	0	1	1	1	1
Rigorous–Intuitive	1	1	1	1	1
Elementary–Advanced	0	1	1	−1	−1
Simple–Complex	1	1	0	0	−1
Clarifying–Mystifying	1	−1	−1	−1	−1
Specific–General	0	0	0	0	0
Concrete–Abstract	1	0	0	0	0
Constructive-Nonconstructive	0	−1	1	1	−1
Practical–Impractical	1	0	1	0	0
Elegant–Ugly	0	1	1	1	−1
Meaningful-Pointless	1	0	0	0	−1
Fruitful–Sterile	1	0	0	1	−1

Note that the fifth proof is worse than proofs (2), (3), and (4).

7.3 The Example of the Primes

In this section we compare four proofs of the infinitude of primes. The mathematics is, in some cases, a little advanced, but the reader can grasp the essential philosophical points by merely 'looking over' the notation.

(1) Euclid's Proof

In Proposition IX 20 of the *Elements,* Euclid points out that if

$$a, \ b, \ c, \ d, \ \ldots, \ z$$

is a finite list of primes and N is their product, then $N+1$ has a prime factor not on the list. Hence no finite list of primes is complete. For example, if we start with the list

$$2, \ 3, \ 5, \ 7, \ 11, \ 13$$

the product $N = 30,030$ and $30,031$ has factor 59.

This proof is at once complete, rigorous, brief, and efficient. Although it is at a suitably elementary level for the theorem, it shows *why* the theorem is true. Note that it does not depend on any logical principles offensive to an intuitionist. On the contrary, it gives a practical way of constructing a prime not on a given list. Since Euclid's proof offers a simple, effective path to the infinite, it is meaningful. Since it allows a number of generalisations (to show, for example, that no list of primes of the form $4m+3$ is complete), it is fruitful. We summarise these comments in the first column in the table below.

(2) Euler's Proof (1737)

If p is a prime, let

$$f(p) = \frac{1}{1 - 1/p} = 1 + \frac{1}{p} + \frac{1}{p^2} + \frac{1}{p^3} + \cdots$$

If N is an integer ≥ 2, let $q(N)$ be the product of the f's for all primes p which are $\leq N$. Then

$$q(N) = 1 + \sum \frac{1}{m} \quad (*)$$

where the sum is taken over all positive integers m that are products of primes $\leq N$. Hence

$$q(N) > 1 + \frac{1}{2} + \frac{1}{3} + \frac{1}{4} + \cdots + \frac{1}{N}$$

Since, as Oresme showed in the Middle Ages, the harmonic series diverges,

$$\lim_{N \to \infty} q(N) = \infty$$

Hence there are infinitely many primes.

This proof, at least as Euler presented it, lacks rigour. In connection with (∗), there is no demonstration of the fact that the product of the geometric series $f(p)$ — for $p = 2, 3, 5, \ldots$ — equals the infinite sum obtained by multiplying their terms in the 'natural' way. Furthermore, the proof is too complex and too advanced for the simple, elementary task of showing the infinitude of primes. On the other hand, it prepares the way for Euler's stronger result that the series of the reciprocals of the primes diverges. In this sense, it is a meaningful and fruitful piece of mathematics.

(3) The Fermat Prime Proof

Let a and k be positive integers with $a > k$. Now

$$
\begin{aligned}
2^{2^a} - 1 &= (2^{2^{a-1}} - 1)(2^{2^{a-1}} + 1) \\
&= (2^{2^{a-2}} - 1)(2^{2^{a-2}} + 1)(2^{2^{a-1}} + 1) \\
&= \ldots \\
&= (2^{2^{a-k}} - 1)(2^{2^{a-k}} + 1)(2^{2^{a-k+1}} + 1) \ldots (2^{2^{a-1}} + 1)
\end{aligned}
$$

Thus if positive integer b is less than a then $2^{2^b} + 1$ divides evenly into $2^{2^a} - 1$. Hence no prime p divides evenly into both $2^{2^b} + 1$ and $2^{2^a} + 1$, lest it divide

$$2^{2^a} + 1 - (2^{2^a} - 1) = 2$$

Now consider the sequence whose general term is

$$2^{2^n} + 1$$

No prime divides more than one of these numbers, and yet they all have prime divisors. Hence there are infinitely many primes.

This proof is unnecessarily complex, tending to mystify more than clarify. It is constructive in the sense that each time we factor a new number in the sequence, we get a new prime, but these numbers increase so rapidly that this is not a very practical way to construct primes. The proof is tricky, but pointlessly so.

(4) The Profinite Topology Proof

In what follows the letters denote nonnegative integers. Where d and e are nonzero, let

$$A = \{r, \quad r + d, \quad r + 2d, \quad r + 3d, \quad \ldots \}$$

and

$$B = \{s, \quad s + e, \quad s + 2e, \quad s + 3e, \quad \ldots \}$$

be arithmetic progressions. Then their intersection is an arithmetic progression (counting the empty set as an arithmetic progression).

For let T be the smallest term they have in common (if there is one). Let $g = \gcd(d, e)$. Then

$$C = \{T, \quad T + de/g, \quad T + 2de/g, \quad \ldots \}$$

is an AP contained in both A and B: $C \subseteq A \cap B$. Moreover, if

$$T = r + md = s + ne$$

and $r + m'd = s + n'e$, so that $s + n'e \in A \cap B$, we have

$$m' = \frac{(r + md - ne) + n'e - r}{d} = \frac{(n' - n)(e/g)}{d/g} + m$$

so that d/g is a factor of $n' - n$, say $n' - n = Qd/g$. Hence

$$s + ne' = s + ne + Qde/g \in C$$

And hence $A \cap B \subseteq C$.

Let us call a set of nonnegative integers *open* iff it is a (possibly infinite) union of arithmetic progressions (or the empty set). Since

$$(\cup A_i) \cap (\cup B_j) = \cup(A_i \cap B_j)$$

it follows that a finite intersection of open sets is open. Hence we have a 'topology'.

Let $K = \cup\{0, \quad p, \quad 2p, \quad 3p, \quad \ldots \}$ where the union is taken over all primes p. Suppose there are only finitely many primes. Then the complement $\overline{K}$ of K is the intersection of finitely many complements of arithmetic progressions. Since the complement of an arithmetic progression is open, it follows that $\overline{K}$ is open (since a finite intersection of open sets is open). But $\overline{K}$ is the set whose only element is 1, and this is not an open set. Contradiction. Hence there are infinitely many primes.

This proof is ridiculously long, advanced and complicated for what it is trying to do. Only a mathematician who loved vocabulary more than thought would make such a mystery of our very specific problem by approaching it from such a general viewpoint. Misplaced topology makes ugly number theory.

We summarise our ratings for the above four proofs in the following table.

SOLUTION: Four Proofs of Prime Infinitude

Dichotomy	Points			
	1	2	3	4
Complete–Incomplete	1	0	1	1
Rigorous–Intuitive	1	−1	1	1
Elementary–Advanced	1	−1	0	−1
Simple–Complex	1	−1	−1	−1
Clarifying–Mystifying	1	0	−1	−1
Specific–General	0	0	0	−1
Concrete–Abstract	0	0	0	0
Constructive-Nonconstructive	1	0	1	−1
Practical–Impractical	1	0	−1	0
Elegant–Ugly	1	0	0	−1
Meaningful-Pointless	1	1	0	−1
Fruitful–Sterile	1	1	−1	−1

The first proof is better than the others.[3]

7.4 Ethical Implications

If mathematical proofs can be evaluated as we have indicated above, there may be some implications for metaphysics and ethics. For example, suppose that one proof is vastly more elegant than another. Every mathematician agrees that the first proof is 'neat', but the second proof is 'clumsy and tedious'. No matter what cultural background is presupposed, the first proof enjoys aesthetic merits entirely lacking in the second. Given that this is the case — and it is the case in some of our examples — it suggests that, *necessarily*, the first proof is more elegant than the second: there is no possible universe in which the first proof is not more elegant than the

---

[3] For examples in more advanced number theory, the reader may wish to consult Anglin's McGill University PhD Thesis 'The Nature of Solutions of Mathematics'.

second. Of course, there are possible universes in which some intelligent creatures *believe* that the second proof is more elegant, but these are universes in which some intelligent creatures are cursed with a twisted sense of elegance. It is simply *perverse* to prefer an obscure, repetitious, ten page derivation of the quadratic formula to the usual five line derivation.

From the fact that there are necessarily true value judgements, it follows that there are also necessarily true ethical judgements. For example, some mathematics students listen to long, rambling proofs (mumbled by the unprepared teacher) only to discover later that the whole topic is covered in a few clearly written paragraphs in one of the standard outlines. Mathematics students with this experience might agree that

> It is not possible that the teacher has a right to inflict such bad proofs on the class.

An argument for this could be made as follows.

> Necessarily, if anything is ugly or inferior, there is a prima facie moral obligation not to propagate it.
> Necessarily, Professor Smith's proof of that theorem is ugly and inferior.
> Hence, necessarily, there is a prima facie moral obligation not to propagate Professor Smith's proof of that theorem.

7.5 Appendix: Hardy's Ratings

In two *Mathematical Gazette* articles (written in 1909 and 1916), G. H. Hardy rates eleven different proofs of the fact that

$$\int_0^\infty \frac{\sin x}{x}\, dx = \frac{\pi}{2}$$

To do this, Hardy assigns 'marks' for difficulties, artificialities, and complexities. Since he thinks it is bad to have too many of these, he claims that the proof with the lowest total is the best. Hardy's marks for difficulties, artificialities, and complexities might show up in our own rating scheme as −1's in the categories 'elegant-ugly', 'meaningful-pointless', and 'simple-complex', respectively.

In this section we give the proof which Hardy judged the best (with only 28 marks), and the proof he judged the worst (with 108 marks). We also give our own ratings for these proofs.[4]

---

[4] The usual contour integral proof of complex analysis received 32 marks, coming in closely behind the winner.

The winning proof goes as follows:

$$\int_0^\infty \frac{\sin x}{x}\, dx = \frac{1}{2} \int_{-\infty}^\infty \frac{\sin x}{x}\, dx$$

$$= \frac{1}{2} \sum_{j=-\infty}^\infty \int_{j\pi}^{(j+1)\pi} \frac{\sin x}{x}\, dx$$

With the substitution $y = x - j\pi$, we obtain

$$\frac{1}{2} \sum_{j=-\infty}^\infty \int_0^\pi \frac{(-1)^j \sin y}{y + j\pi}\, dy = \frac{1}{2} \int_0^\pi \sin y \sum_{j=-\infty}^\infty \frac{(-1)^j}{y + j\pi}\, dy$$

$$= \frac{1}{2} \int_0^\pi \sin y \csc y \, dy = \frac{1}{2} \int_0^\pi 1 \, dy = \frac{\pi}{2}$$

using a cosecant series that is usually derived from complex analysis (using, say, the Mittag-Leffler Expansion Theorem). Hardy gives 15 marks for the switch of the summation and integral signs, 4 for the use of the cosecant series, and 9 for the rest, giving the total of 28.

In terms of our own system, the proof is complete and rigorous, but it is, perhaps, a trifle too advanced — on account of its use of the cosecant series. The proof is undoubtably short and neat, and, although we may want to look for a way of avoiding the use of the cosecant series, we have to admit that the series is used in a very clever way, even adding to the elegance of the proof. The proof does not make any attempt to explain *why* the theorem is true. We might rate it as follows:

SOLUTION: Hardy's Favourite

Dichotomy	Points	Reasons
Complete–Incomplete	0	
Rigorous–Intuitive	1	everything tied together
Elementary–Advanced	−1	cosecant series
Simple–Complex	1	a slight twist
Clarifying–Mystifying	0	
Specific–General	1	right level for result
Concrete–Abstract	0	
Constructive-Nonconstructive	1	steps numerically accessible
Practical–Impractical	0	
Elegant–Ugly	1	cosecant series
Meaningful-Pointless	0	
Fruitful–Sterile	1	technique generalises

Hardy's booby prize goes to a rather lengthy proof due to Nanson. It is not unentirely unredeemable — as Hardy notes, the method can be used to solve other problems — but the reader will probably not want to include it in his or her textbook on analysis. It begins with an entirely unmotivated definition:

$$u(a) = \int_0^\infty \frac{a \cos cx}{a^2 + x^2} \, dx$$

where a and c are positive. Two integrations by parts (on the trig functions) yield

$$u(a) = \frac{2a}{c^2} \int_0^\infty \cos cx \frac{a^2 - 3x^2}{(a^2 + x^2)^3} \, dx$$

Two differentiations with respect to a, across the integral sign (this having to be justified), give

$$\frac{d^2 u}{da^2} = 2a \int_0^\infty \cos cx \frac{a^2 - 3x^2}{(a^2 + x^2)^3} \, dx$$

and hence

$$\frac{d^2 u}{da^2} = c^2 u$$

Nanson then notes that the substitution $x = ay$ shows that u is a function of ac, so that, applying the theory of differential equations, u has the form $Ae^{ac} + Be^{-ac}$. That $A = 0$ is proved by making $c \to \infty$, and observing that

$$|u| < \int_0^\infty \frac{a}{a^2 + x^2} \, dx = \frac{\pi}{2}$$

That $B = \frac{\pi}{2}$ is proved by making $c = 0$. (Here we assume the continuity of the integral.) This gives

$$u = \frac{\pi}{2}e^{-ca}$$

We then have

$$\int_0^\infty \frac{a\sin mx}{x(a^2 + x^2)}\, dx = \int_0^\infty \frac{a}{a^2 + x^2} \int_0^m \cos cx \, dc \; dx$$

$$= \int_0^m \int_0^\infty \frac{a\cos cx}{a^2 + x^2}\, dx \; dc = \int_0^m u \, dc$$

$$= \int_0^m \frac{\pi}{2}e^{-ca}\, dc = \frac{\pi}{2a}(1 - e^{-am})$$

Differentiating twice with respect to m (across the integral sign on the left-hand side), we get

$$\int_0^\infty \frac{x\sin mx}{a^2 + x^2}\, dx = \frac{\pi}{2}e^{-am}$$

Hardy says, 'this should strictly involve 40, and I cannot reduce the number to less than 30, as the second differentiation is none to easy to justify'. Adding the above formula to

$$a\int_0^\infty \frac{a\sin mx}{x(a^2 + x^2)}\, dx = a\frac{\pi}{2a}(1 - e^{-am})$$

we obtain

$$\int_0^\infty \frac{\sin mx}{x}\, dx = \frac{\pi}{2}$$

and the result follows.

On our rating scheme, Nanson's proof deserves bad marks for being too advanced (using differential equations) and too long. It is also repugnantly mystifying. However, as Hardy notes, Nanson's method can be used to get other results, so we could at least award a point for fruitfulness.

SOLUTION: Hardy's Booby Prize

Dichotomy	Points	Reasons
Complete–Incomplete	0	
Rigorous–Intuitive	0	
Elementary–Advanced	−1	differential equations
Simple–Complex	−1	bizarre complication
Clarifying–Mystifying	−1	confusing tricks
Specific–General	0	
Concrete–Abstract	0	
Constructive-Nonconstructive	0	
Practical–Impractical	0	
Elegant–Ugly	−1	far too long
Meaningful-Pointless	0	
Fruitful–Sterile	1	technique generalises

Chapter 8

Mathematics and History

Some people think that history is a purely objective, ideology-free enterprise that appears, not in popular textbooks or newspapers, but only in specialised academic journals. History is the property of an elite, and someone who has not mastered strange languages does not belong. In this chapter I am going to discuss something else. I am going to discuss something that belongs to the public, and something that is shot through with ideological presuppositions, and something that raises interesting philosophical questions.

8.1 A Caricature

We begin this chapter with a caricature of the typical history of mathematics textbook. This caricature will draw attention to some of the philosophical presuppositions which often underlie these books. We shall then discuss these presuppositions.

Mathematics epitomises Reason. It began in Egypt and Mesopotamia. However, it really began in Greece, because that is where *pure* mathematics began, and pure mathematics is better than applied mathematics, because pure Reason is better than impure Reason.

The great Greek mathematicians were Eudoxus, Apollonius, Archimedes, and Hypatia. Hypatia did little compared to Archimedes, but Hypatia was the only woman who did mathematics, and we are sure that her true greatness was masked by male chauvinism.

In spite of their preference for geometry, and their rejection of motion in mathematics, the Greeks were wonderful. Unfortunately, superstition and ignorance made a comeback when Hypatia (400 AD) 'was hacked to pieces by Bishop Cyril'.[1] For one thousand years, no one in Western Europe did any mathematics.

Meanwhile, the Arabs were developing algebra. Although he could not prove the Theorem of Pythagoras for non-isosceles right triangles, Al-Khwarizmi was a magnificent algebraist. He once found two solutions to a quadratic equation, and he used three different values for π.

In the sixteenth century, Europe rebelled against the Church, and Reason and Happiness returned. Taking up where Hypatia had left off, Newton and Leibniz invented calculus, introducing motion into mathematics. Oddly, they did this independently. Therefore each of them deserves all the glory and praise for creating calculus.

In the nineteenth century, Reason came into her own. Calculus became rigorous (by forgetting about motion), and the infinite was reduced to the finite.

Unfortunately, we cannot tell you much about the twentieth century because it would take all our time and energy just to find out ourselves what is happening. We do know, however, that the Four Colour Map Problem was solved by a computer.

Then there was the extraordinary mathematician X. She was born in Ruritania, and all properly patriotic Ruritanians are justly proud of her. One of her parents died when she was very young, and she showed marvelous mathematical abilities at age 3. X's remaining parent wanted her to be a plumber, but X persevered in her mathematical researches until she was broke and unemployed. Nobody offered her a university position, because no one was able to understand her proof of the theorem that there is no 4 by 4 magic square whose entries are the squares of the first 16 positive integers. Fortunately, there were no bishops around to hack her to pieces.

Thus ends the caricature. It is meant to raise some questions about the nature of the history of mathematics, and, in what follows, we deal with eighteen such questions, suggesting new ways of writing a history of mathematics.

---

[1] G. Toussaint, 'A New Look at Euclid's Second Proposition', *The Mathematical Intelligencer*, 15 (3) (1993), 17.

8.2 Is Math Good?

Should the Historian Write as though Mathematics is always a Good Thing?

Often a person writes a history of mathematics because he or she loves mathematics. William Dunham, for one, does not hide his enthusiasm. His *Journey through Genius* begins by informing us that Bertrand Russell's desire to know more mathematics was strong enough to prevent him from committing suicide.[2] Dunham then compares the great theorems to the plays of Shakespeare, and the paintings of Van Gogh. Here is beauty well worth living for!

A history of mathematics written by someone who hated the subject would make dreary reading. Beside the various advantages of an enthusiastic author, however, we must place one of the drawbacks. Just as a deeply pro-Catholic writer is going to have trouble giving an unbiased account of Luther, so a deeply pro-mathematics historian is going to have trouble giving an unbiased account of, say, the shortage, in a certain century, of university positions for mathematicians. The pro-Catholic author will tend to make martyrs out of Catholics who suffered at the hands of Protestants, and the pro-mathematics author will tend to glamorise or pity mathematicians who could not find decent jobs.

According to Aristotle, humans are characterised by rationality. With this as an assumption, it is not hard to construct arguments for the conclusion that any advance in Reason is a good thing for humanity. From this conclusion, moreover, it is not far to Neo-Platonic religions in which Reason is a god, and mathematical activity is liturgy.

This is not the place to settle the matter of the relative importance and goodness of mathematics, but it is the place to note a couple of examples where the exaltation of mathematics can introduce bias into histories of mathematics.

The first is the case of China. Those who put a high value on mathematics tend to judge a culture by the number of theorems it proved. This leads to an unrealistic view of cultures, such as the ancient Chinese, which did not produce many theorems, but which did produce excellent poetry and philosophy. A properly balanced history of mathematics would recognise the importance of many forms of intellectual activity — including the creation of myths, the development of law, and the pursuit of truth in theology.

A second example of bias is the case of Hypatia, a fifth century pagan mathematician who was killed by a street mob in Alexandria. According to

---

[2] W. Dunham, *Journey through Genius* (New York: John Wiley, 1990), v.

Socrates Scholasticus (380–450), in Chapter 15 of Book VII of his *History of the Church,* the mob was led by a 'Christian' called Peter. An anti-mathematics historian might describe Hypatia's death as the removal of an arrogant, upper class reactionary who stood in the way of the new Christian society. The pro-mathematics historian, however, invariably beatifies Hypatia, lamenting her death as a sign of the decline of Reason in the West. If you like labels, Hypatia was a 'martyr for science', but she was also a 'benighted polytheist', an 'ivory-tower elitist', and a 'rich, pampered parasite'. Believers in Reason feel that Hypatia's modest work in Diophantine equations was more important than the Alexandrians' need to reject upper class paganism, but this feeling is a value judgement, not an objective fact about the fifth century.

8.3 Individualism

Should a History of Mathematics Revolve around Individuals and their Private Lives?

In Boyer and Merzbach's *A History of Mathematics,* ten out of twenty-eight chapters are named after individual mathematicians. In the fifth edition of *An Introduction to the History of Mathematics,* Eves includes pictures of Pythagoras and Archimedes, even though we have no idea what they really looked like. In the Preface of *Journey through Genius,* Dunham asserts that an understanding of the private lives of mathematicians 'can only enhance an appreciation of their work'. In the Preface of *The History of Mathematics,* Burton writes:

> Considerable prominence has been assigned to the lives of the
> people responsible for progress in the mathematical enterprise.
> In emphasizing the biographical element, I can say only that
> there is no sphere in which individuals count for more than [in]
> the intellectual life.

Most histories of mathematics are individualist, as opposed to, say, Marxist. The historian arranges the material in terms of individuals, and goes to much trouble to ensure that the right individual receives the right amount of praise for the right theorem. Individualist historians tend to be thrown off balance when a theorem does not have a unique, nameable 'first discoverer'.

One should note that there are alternatives. There is no reason why one could not write a history of mathematics from a communitarian point of view. Instead of singling out the individual responsible for the theorem, one could single out the technological capacities or social needs which were responsible for it. Instead of glorifying the lucky person who happened to be the first to get the proof, one could glorify the ethical ideals of the community which led it to educate people in such a way that this proof was inevitably found.

Individualist historians of mathematics like to repeat anecdotes about Euclid — even though they know that there is no factual basis for these anecdotes. They do not tell you whether the *Elements* is an expression of the upper class elitism we detect in Plato's dialogues, and they do not address the issue of whether the purity of Euclid's proofs proclaims a contempt for manual work (this being performed by the slaves).

If an individualist is someone who believes that an individual, by his or her free choice, can make a significant difference to history, then I am an individualist. However, if an individualist is someone who believes that praise should be lavished on famous persons, then I am not. For most of the work is done, not by the individual mathematician who first proves the theorem, but by a whole collection of supporting persons. Without God or society, without teacher or spouse, the famous individual would accomplish nothing. Newton wrote:

> If I have seen farther than others, it is because I have stood on the shoulders of giants.

8.4 Nationalism

Should a History of Mathematics be Organised in Terms of Nations or Races?

D. E. Smith divides the first volume of his *History of Mathematics* into ten chronologically determined chapters. These chapters are subdivided into sixty-seven sections, and thirty-two of these sections are named after nations (e.g. Egypt, England). Smith thus follows the customary practice of organising the history of mathematics by nationality. In his Preface, Smith asserts:

> While it is evident that no race or country has any monopoly of genius, and while the limits of successive countries are only artificial boundaries with no significance in the creation of the masterpieces of science, nevertheless linguistic and racial influences tend to develop tastes in mathematics as they do in art and in letters.

He continues:

> In this treatment of the subject an attempt has been made to
> seek out the causes of the advance or the retardation of math-
> ematics in different countries and with different races.

One example of Smith's 'nationalistic analysis' occurs at the beginning of
the 'Germany' section in the chapter on the sixteenth century:

> The mathematics of Germany was Gothic, unpolished, but vir-
> ile; the mathematics of France was Renaissance, polished, but
> generally weak.

Does this rather vague — and potentially offensive — comment throw light
on sixteenth century mathematics?

There are two reasons for asking historians of mathematics to stop organising their thought along nationalistic or racial lines. The first is ethical.
Patriotism and racism lead to war. It is not good for readers to have their
patriotic blood heated up over the fact that the Germans beat the French
to the discovery of, say, a proof of the Law of Quadratic Reciprocity.

The second reason for asking historians to stop describing mathematical
activity in terms of nations is that mathematics is a universal enterprise,
with intellectuals from every corner of the globe pursuing a common goal,
a goal which transcends political and racial boundaries. Admittedly, some
mathematical research is tied to the military ambitions of particular nations, and, in such cases, it may be important historically if one country,
rather than another, is the first discoverer of some theorem. Typically,
however, mathematical research furnishes one of the best examples of international cooperation.

Smith's treatment of Fibonacci (1170–1250) is interesting in this connection. Smith's 'nationalistic analysis' leads him to remark that the work
of Fibonacci, an Italian, was beyond the competence of any professor in
Paris.

What needs to be told is quite different. The scholars of medieval Europe were not tied to their particular countries. They spoke a common
language (Latin), and they travelled all over the continent. At that time,
the sharp social boundaries were not national or racial, but religious. The
point which Smith's nationalism obscures is that Fibonacci was the best
mathematician, not just in Italy or France, but in the whole European
scholarly community. And one should also mention that Fibonacci was
educated in Africa.

David Hilbert was a staunch internationalist. Addressing the International Congress in Bologna in 1928, he asserted that

> all limits, especially national ones, are contrary to the nature of
> mathematics. It is a complete misunderstanding of our science

to construct differences according to peoples and races, and the reasons for which this has been done are very shabby ones.[3]

8.5 Women

How Should Historians Tackle the Scarcity of Women in Mathematics?

Men and women are equal intellectually.[4] Apparent differences between male and female mathematical ability are due to social factors such as cultural systems in which men take all the educational opportunities for themselves. For whatever reason, however, the fact remains that, prior to 1900, there were fewer than a dozen famous women mathematicians.

What should the historian do about this fact? There are at least three things to be done, and one thing to be avoided. The thing to be avoided is a false exaggeration of the role of women in the history of mathematics.

No doubt with good intentions, David Burton exaggerates the importance of certain women mathematicians. In *The History of Mathematics,* he tells us that Theano, the wife of Pythagoras, was

> a remarkably able mathematician, who not only inspired him during the latter years of his life, but continued to promulgate his system of thought after his death.[5]

There is no evidence at all for these assertions. We know that Theano was herself a Pythagorean, but we have no testimony about her mathematical abilities, or about her influence on Pythagoras. Indeed, we do not even know whether Pythagoras himself was a 'remarkably able mathematician'.[6]

Burton also exaggerates the importance of Hypatia. He assures us that 'with the death of Hypatia, the long and glorious history of Greek mathematics was at an end'.[7] This is going too far. The last first-rate ancient Greek mathematician was Pappus, who lived a century before Hypatia. The last second-rate ancient Greek mathematician was Proclus, who died seventy years after Hypatia. Depending on one's standards of glory, the 'long and glorious history of Greek mathematics' ended with either Pappus or Proclus — but not with Hypatia.

Attempts to play up the role of women in mathematics distort history and patronise women. Ann Hibner Koblitz writes:

---

[3] C. Reid, *Hilbert,* p. 188.

[4] *Galatians* 3:28.

[5] D. M. Burton, *The History of Mathematics,* 2nd edn., p. 92.

[6] As far as the Theorem of Pythagoras is concerned, we not do know whether Pythagoras was to first to prove it, but we do know that he was not the first to discover it.

[7] Ibid., p. 242.

> It is unfortunately the custom among some feminist writers to-
> day to exaggerate the past achievements of women, to claim
> that this or that famous woman was the equal of or better than
> the greatest males in the same field. Since it is often impossi-
> ble to prove these claims, as in many cases they are not true,
> the net effect of their efforts is to belittle the real, important
> achievements of the women they wish to aggrandize.[8]

If the historian is not to falsify the importance of the few women mathematicians, is there some other way of coping with the results of male
chauvinism? There are at least three ways.

First, the historian can draw attention to the fact that there were some
great woman mathematicians. David Burton is not exaggerating in the
extended treatment he gives to Emmy Noether (1882–1935).

Second, the historian can challenge the veneration which histories of
mathematics usually accord to mathematics. Perhaps the reason there are,
even today, few women in mathematics is that women realise that careers
in mathematics are stressful, low-paying, and socially useless. Only a man
would be stupid enough to want one.

Third, the historian can abjure the individualism so prevalent in histories of mathematics. Why does the praise go only to the person who
actually completes the solution of the mathematical problem? Could he
have solved it without the support of his wife or lover or mother? The
cultural output of a society depends on the effort and creativity of the
whole population, and it is inaccurate to attribute that output to the few
individuals who merely add the finishing touches.

8.6 Minorities

How Should Historians be Fair to Minorities?
In his book *A History of Mathematics,* Victor Katz has a section on Ben-
jamin Banneker (1731–1806). Banneker was not an original mathematician.
He was not really up on the research of his times. But he was black, and,
presumably, Katz has included him simply in order to show fairness to a
minority.

Needless to say, fairness is a virtue, and it is a wicked historian who
deliberately ignores a highly competent mathematician just because that
mathematician belongs to a minority. However, it may be possible to go too
far in the other direction. Suppose an historian writes up a glowing report
on the first mentally challenged Canadian to pass a course in algebra. Is

---

[8] A. H. Koblitz, *A Convergence of Lives* (Boston: Birkhäuser, 1983), p. 269.

this a wonderful example of good liberal principles, or is this just wasting our time with trivia? Is this conscientious fairness, or is this an unintended mockery of the mentally challenged? Is the author being 'thorough' in his treatment, or is he merely being patronising? Is the author interested in the mentally challenged, or is the author just interested in appearing to be politically correct? One would reasonably expect to find an account of the first mentally challenged Canadian to pass a course in algebra in a book devoted to the history of the mentally challenged in Canada, but it would seem odd to find such an item in a general history of mathematics book.

8.7 Charity

How Charitable Should Historians Be?

In the *Disquisitiones Arithmeticae* (1801), Gauss writes:

> it is clear from elementary considerations that any composite number can be resolved into prime factors, but it is tacitly supposed and generally without proof that this cannot be done in many various ways.[9]

Gauss then gives what is often taken to be the first explicit proof that factorisation is unique.

Certain authors, however, claim that it was, not Gauss, but Euclid (or one of the ancient Pythagoreans) who first proved this 'Fundamental Theorem of Arithmetic'. T. L. Heath summarises Proposition IX 14 of the *Elements* by saying, 'in other words a number can be resolved into prime factors in only one way'.[10] H. N. Shapiro agrees: 'Proposition 14 is essentially the Unique Factoristion Theorem'.[11]

A quick look at Proposition IX 14 reveals that, indeed, 'what Euclid was getting at' is what we call the Fundamental Theorem of Arithmetic. However, a closer look at that Proposition also reveals that he only proved it for the case of square-free integers. His proof does not work for integers, such as 50, which have nontrivial square factors. Moreover, the proof of Proposition IX 14 is based, ultimately, on Proposition VII 20, and the proof of Proposition VII 20 is faulty. This rather uncharitable evaluation of Proposition IX 14 is confirmed by the fact that, in the proof of Proposition IX 36, Euclid does not cite Proposition IX 14 to establish the unique

---

[9] C. F. Gauss, *Disquisitiones Arithmeticae*, p. 6.

[10] Euclid, *The Elements,* 2nd edn., vol. 2, p. 403.

[11] H. N. Shapiro, *Introduction to the Theory of Numbers* (New York: John Wiley, 1983), p. 43. See also p. 817 of M. Kline's *Mathematical Thought from Ancient to Modern Times,* and p. 107 of the 5th edn. of H. Eves *An Introduction to the History of Mathematics.* Both Kline and Eves agree with Heath and Shapiro.

factorisation of integers of the form

$$2^{n-1}(2^n - 1)$$

but gives a wholly independent proof of the unique factorisation of such integers (one which also depends on Proposition VII 20).

In the light of this, it seems that Heath and Shapiro are too charitable. The *Elements* does not contain a statement or a proof of the Fundamental Theorem of Arithmetic.

How charitable should one be? On the one hand, one must grant that an author may have had more in mind than his mathematical notation allowed him to express. On the other hand, it is not responsible scholarship to ascribe to an ancient writer a level of generality for which there is no hard evidence.

As a guideline, I propose this: do not be charitable to earlier mathematicians at the expense of later ones. Unless it is really in the *Elements*, you have no right to praise Euclid by robbing Gauss.

8.8 Chronological Organisation

Should the History of Mathematics be Divided into Chronological Periods?

W. W. Rouse Ball organises *A Short Account of the History of Mathematics* in terms of three chronological segments: the Greek period (600 BC to 641 AD), the medieval and Renaissance period (641 to 1637), and the modern period. Similar chronological divisions are found in many histories of mathematics — although some accord a period to the ancient Egyptians and Mesopotamians, and some split the Renaissance off from the Middle Ages. A thousand years from now (if the world has not ended), historians may give the year 1950 as the end of the 'European Calculus Period' and the beginning of the 'International Computer Period'.

It is useful to divide history into chronological periods, but one must be careful not to over-simplify. The way some historians tell it, all the really worthwhile mathematics was done in the Greek and modern periods. Nothing at all occurred in the Middle Ages. Hollingdale, for example, entitles his short chapter on medieval mathematics 'The Long Interlude'.

Perhaps someone should show Hollingdale a list of some impressive medieval achievements.

(a) The Chinese developed 'Horner's method'.

(b) Chinese or Indian mathematicians introduced negative numbers.

(c) Al-Mu'taman (1085) gave the first known proof of Ceva's theorem.

(d) Fibonacci (1225) found an integer solution to the system

$$x + y + z + x^2 = w^2$$
$$w^2 + y^2 = u^2$$
$$u^2 + z^2 = v^2$$

(e) Oresme (1350) gave the first proof of the divergence of the harmonic series.

(f) Albert of Saxony showed that the members of an infinite set can be put in one-to-one correspondence with the members of a proper subset of it.[12]

8.9 Dates

How Should Historians Date Mathematical Discoveries?

It is sometimes difficult to give 'the' date of a discovery. Problems such as the following arise.

(a) Smith published a clear statement and proof of Theorem T in year Y. However, something 'essentially the same as' Smith's result had appeared in an obscure journal in year Y − 20.

(b) Theorem T appeared in print for the first time in year Y. However, the clever and honest Smith claimed he was 'in possession of it' in the year Y − 20.

(c) In year Y, Jones published a proof of Theorem T, claiming that he was only presenting the ideas found in Smith's notes written in year Y − 20. However, Jones was the only mathematician who was ever able to make any sense out of those notes.

(d) Theorem T was 'substantially proved' in year Y − 20. Later it was realised that many details still needed to be filled in, and this was done by a graduate student in year Y.

(e) Smith discovered a proof of theorem T in year Y − 20, but failed to realise he had done so — until he recognised what was 'essentially' his own work in a paper given by Jones in year Y.

(f) Smith published a result in year Y − 20, but never said anything to indicate he was aware of its almost immediate corollary, Theorem T. This corollary was not stated until year Y, when it was given by Jones. Jones ascribed it to Smith 'who must have been aware of it'.

---

[12] J. H. Ewing, *Numbers,* trans. H. L. S. Orde (New York: Springer-Verlag, 1990), pp. 355-6.

Cases such as the above do occur. Some of them arose in connection with the discovery of hyperbolic geometry.

The wise historian will not try to decide whether Theorem T was discovered in year Y or in year $Y - 20$. The wise historian realises that discoveries often occur, not on single dates, but over a period of years or decades. Just as the Luther Dean takes several minutes to regain consciousness when he wakes up on Monday afternoon, so it takes mathematicians, as a group, several years, or even decades, to become fully aware of their deeper, more important results. At first they get a mere glimpse of the theorem. Then they discover some new interconnection. Next someone publishes the key insight. Then someone else irons out the details, making important simplifications. Next someone finds the most apposite generalisation. Finally, a textbook writer gives what is, in fact, the first clear and complete presentation of the matter. The essential ideas are grasped around the year $Y - 20$, but the first rigorous proof appears only in year Y. In cases such as these, the historian must abandon any attempt at individualism, and admit that the proof was discovered by Smith and Jones (and Brown too), and in all the years between $Y - 20$ and Y.

8.10 Calculating Devices

Should Historians Ignore Calculating Devices?

By 'calculating devices', I mean algorithms for calculating, as well as machines used to implement them. In this sense, calculating devices include the abacus, Horner's method, the long division procedure, slide rules, and computer networks.

Historians of mathematics sometimes neglect calculating devices. Burton, for example, says virtually nothing about computers in *The History of Mathematics*. The word 'computer' does not appear in the index of his book.

Perhaps the reason for this is a belief that calculating devices have little impact on the progress of 'real' mathematics, and it may be useful to give some counterexamples.

(a) It was only in the context of a set of algorithms for positional arithmetic that negative numbers could be introduced. The purely theoretical treatment of number given by the ancient Greeks was not enough.

(b) It was only by comparing large numbers of functions, slopes and areas that the calculus was developed. This required the calculation of many 'tables of values', a task that would not have been feasible without the decimal system, and the use of logarithms.

(c) Much contemporary number theory is done by means of computer interaction. Before the computer, we knew of only 12 perfect numbers; now we know of over 30.

(d) It is only thanks to the computer that the Four Colour Theorem was finally proved.

From the above examples, we see that, even in pure mathematics, it may happen that a problem will not be solved without the help of sufficiently sophisticated computational technology.

8.11 Astronomy

Should Historians include Astronomy with Mathematics?

The same historians who praise ancient Greek mathematics for its otherworldliness also praise seventeenth century mathematics for its practical applications. In doing the former, they ignore ancient Greek astronomy. In doing the latter, they give a great deal of attention to seventeenth century astronomy. Ancient Greek astronomy does not count as mathematics, but seventeenth century astronomy does!

Why is this so? Perhaps some Hegelian theory is at fault. The mundanity of ancient Egyptian mathematics has to be countered by an otherworldliness in ancient Greek mathematics. The transcendent mathematics of the Age of Faith has to yield to an immanent mathematics in the Age of Empiricism. When the pendulum swings to the ideal, we ignore astronomy. When it swings to the practical, we embrace it.

To be complete, a history of mathematics has to include both pure and applied mathematics, and it has to say something about the interaction between them. Ancient Greek astronomy should be included because it gave rise to trigonometry. Seventeenth century astronomy should be included because it was a neat application of theoretical work on the conics.

8.12 Transcendence

Should Historians Portray Mathematics as Transcendent?

Some historians exalt the purity, rigour, or perfection of mathematics, presenting the mathematician as an otherworldly mystic.

For example, in *Mathematics and Its History,* John Stillwell repeats the anecdote about Euclid's reply to a student who asked about the job market for mathematics graduates. Euclid called his slave and said: 'Give him a coin if he must profit from what he learns'.[13] The idea is that the mathematician is so far above money that he will not even touch it.

Historians of mathematics sometimes take this attitude for their own. If the famous mathematician is poor and unemployed, they glamorise his otherworldliness, but if he is shrewd and rich, they are too embarrassed to mention it. We hear about 'poor, young Abel', but not about 'rich, old Gauss'.[14] If the mathematician devotes himself to some ultra-otiose topic, the historians long to share the details, but if the mathematician is a sensible sort, working in actuarial science, the historians ignore him. The only financial mathematics ever mentioned consists of some quaint, medieval inheritance puzzles.

The fondness of some historians for the transcendent shows up in their treatment of kinematics. According to Plato, the transcendent world is changeless, and mathematics involving motion is not real mathematics. Historians who agree with Plato present geometry rather than astronomy as the 'heart' of Greek mathematics (although the Greeks themselves may have seen it differently). They explain that Weierstrass did calculus a great service by purging it of the concept of motion.

8.13 Rigour

Should Historians Idolise Rigour?

There are two views of rigour and the historian should be careful not to get too carried away with either.

(a) Rigour is the essence of mathematics. If it is not a rigorous deduction, it is not real mathematics. This position is upheld by Hollingdale when he quotes G. H. Hardy:

> The Greeks were the first mathematicians who are still "real" to us today. Oriental [Egyptian or Mesopotamian] mathematics may be an interesting curiosity, but Greek mathematics is the real thing.[15]

(b) Rigour is the fossilization of mathematics. Mathematics progresses, not via deduction, but via experimental science and artistic insight. Math-

---

[13] J. Stillwell, *Mathematics and Its History,* p. 25.

[14] Ewing, *Numbers,* p. 111.

[15] S. Hollingdale, *Makers of Mathematics* (London: Penguin Books, 1989), p. 12.

ematics is not a careful march down a well-cleared highway, but a journey into a strange wilderness, where the explorers often get lost. 'Rigour' is a signal to the historian that the maps have been made, and the explorers have gone elsewhere.

At the moment view (b) is in the ascendant, so the historian might be wise to recall the fact that some mathematicians, such as Cauchy, held view (a), and did at least some of their mathematics accordingly.

8.14 Details

How should the historian handle details?

Some journals, for example, like nothing better than to publish articles on the notational changes in the second draft of Professor Smith's proof that there are 25 primes less than 100.

Yes, good history does require the reporting of details, but there is much to be said for having a larger picture too. If one's details do not help develop a larger picture, or if there is no larger picture to be developed, then perhaps one should consider early retirement.

Is the historian of mathematics supposed to take pride in having written 100 articles of infinitesimal narrowness, or is the real achievement the communication of a vision of the whole history of the subject, the communication of an apppreciation, as well as an understanding, of its trials and triumphs?

8.15 Fantasy Mathematics

Should Historians Report on Mere Fantasy Creations?

We should be nonplussed if a history of explorations included, not only accounts of Columbus discovering America, and Scott journeying to Antarctica, but also accounts of Dorothy discovering Oz, and Frodo journeying to Mordor. From a Platonist point of view, the history of mathematics is a history of exploration, and an exploration of independently existing mathematical structures (such as the Euclidean plane) must be distinguished from a fantasy creation of merely mental structures (such as, say, the hyperbolic plane). If, for example, the Continuum Hypothesis is true, any mathematics based on the opposite assumption is mere science fiction, and no more belongs in a history of mathematics than the Land of Oz belongs in a history of exploration.

To this one might object that if many people believed in Oz, and if Geography Departments regularly taught courses on it, then even a serious history of exploration should include a chapter on Oz. Similarly, given that Mathematics Departments teach courses in hyperbolic geometry, it is necessary to include that subject in a history of mathematics, even if there really is no hyperbolic plane to explore.

As in the case of rigour, one can talk about a distinction between 'real' mathematics and Brand X. It is, of course, the 'real' mathematics that should attract the historian's attention — but then who is to decide what is 'real'? If historians, acting on some value judgement, try to weed out the 'darnel', they may 'pull up the wheat with it'.[16]

8.16 Epic or Comedy?

Is the History of Mathematics an Epic or a Comedy?

Historians sometimes present the progress of mathematics as a march of knowledge against evil ignorance. Every discovery is an important component in the overall victory of Reason. Historians of this ilk do not tell jokes.

As an historical fact, the history of mathematics is filled with laughter. Mathematicians are reputed to be intelligent and rational, but their work is frequently mixed with error, omission, confusion, or even superstition. It would be easy to write of history of mathematics that was both accurate and funny. Here are a few examples.

(a) In 1799, Gauss published a proof of the Fundamental Theorem of Algebra. He prefaced this proof with an aculeate criticism of the previous 'proofs' of that theorem, showing how they were all wrong. Some earlier mathematicians, for example, had taken to calling the Fundamental Theorem 'D'Alembert's Theorem', because they mistakenly thought D'Alembert had proved it, in 1746. Evidently these mathematicians were better at praising D'Alembert's work than understanding it. (Ha! Ha!) The funny part, however, is that Gauss's own proof was flawed.[17]

(b) The early workers in calculus reasoned by analogy. They took principles which apply in finite cases, and applied them to examples involving the infinite. George Berkeley laughed at them:

> He who can digest a second or third fluxion, a second or third difference, need not, methinks, be squeamish about any point

---

[16] *Matthew* 13:29.
[17] Stillwell, *Mathematics and Its History*, pp. 195-200.

in Divinity.[18]

About a hundred and fifty years later, calculus was finally put on a rigorous footing, and mathematicians congratulated themselves that they were, at last, above the jibes of Berkeley. The funny part is that, in 1966, Abraham Robinson showed that the infinitesimals which Berkeley had ridiculed were basically sound.

(c) Euclid intended to give a rigorous treatment of geometry but he goofed in his very first demonstration. He forgot to add an axiom to ensure that the circumferences of two overlapping circles actually meet in a point (rather than just passing through each other without touching). In 1899, David Hilbert presented a 'rigorous' re-working of Euclid — but made the same mistake. It was only in the French translation of Hilbert's book that he belatedly added the necessary axiom.[19]

(d) In 1968, R. Kershner claimed that he had established a complete list of all possible ways of tiling a plane with congruent, convex pentagons. He published his result in the prestigious *American Mathematical Monthly*. Sure enough, in 1975, someone called R. E. James produced a tiling Kershner had missed. The funny part, however, is that the expert on the subject turned out to be a San Diego housewife, Marjorie Rice, who had never been to university, but who found four tilings that had eluded Kershner and James both.[20] As Kersher himself had noted, in a book he published way back in 1950, established 'theorems' are sometimes nonsense.

There are other examples too: Cauchy's misadventures with uniform convergence, Kronecker's silly denial of the infinite, Lambert's failure to realise he had created a new geometry, and so on. Usually these examples are presented as minor oversights, later corrected by the 'inevitable increase in rigour'. However, presented as the ludicrous lapses they really were, these 'oversights' make entertaining instruction, and give a key insight into the nature of mathematical activity.

8.17 Sex

Should an Historian of Mathematics Write about Sex?

Some historians avoid the topic altogether, but most textbook writers do say something about sex. Victor Katz is an example. He goes out of

---

[18] C. B. Boyer, and U. C. Merzbach, *A History of Mathematics,* 2nd edn. (New York: John Wiley, 1989), p. 465.

[19] D. Hilbert, *Foundations of Geometry,* p. 25.

[20] D. Schattschneider, 'In Praise of Amateurs', in *The Mathematical Gardner,* pp. 140-66. The tiling problem was still open in 1993.

his way to tell us that Qin Jiushao (1202–1261) had a house with a 'series of rooms for lodging beautiful female musicians and singers'.[21] He informs us that the word 'sine' comes from a Latin word for breast. He quotes one of Mahavira's more salacious mathematical word problems, and he alludes to the incestuous-sounding poetry in the mathematics book written by Bhaskara for his daughter Lilavati. From Katz we learn that it was al-Samaw'al (1125–1180) who perfected our modern long division algorithm — as well as composing a collection of erotic stories called *The Companion's Promenade in the Garden of Love.* Nor is Katz is afraid to tell us that Alan Turing's mathematical career ended when he was arrested for being a practising homosexual.

John Stillwell is another historian who writes about sex in the lives of famous mathematicians. In his book *Mathematics and Its History,* he reports that Descartes had a daughter, Francine, by his mistress, Helen, and he also mentions the fact that Kurt Gödel married 'a dancer at a nightclub in Vienna'.[22]

One might question the use of these details. Do they help us understand the great mathematicians? Are they related in any way to the actual mathematics? Could they have an 'untoward' influence on people who are immature?

In defense of writers such as Katz and Stillwell, one might make the following points:

(1) Erotic details can provide a key to understanding the cultural background of the lives, and work, of the famous mathematicians.

(2) Erotic details can provide important evidence for theories correlating sex with academic productivity.

(3) Truth has greater value than religious or political correctness.

(4) Freedom of speech, and academic freedom, are essential components of a wholly good society.

(5) The best way to help immature people become mature is to treat them as if they were already mature.

---

[21] V. J. Katz, *A History of Mathematics,* p. 189.
[22] J. Stillwell, *Mathematics and Its History,* p. 330.

8.18 Religion

Is the History of Mathematics a War against Religion?

Auguste Comte (1798–1857) maintained that religion is an enemy of learning, and, in particular, an enemy of mathematics. Some historians of mathematics agree. For example, Burton writes:

> a new movement developed in Alexandria, and also in many other parts of the empire, which was to accelerate the demise of Greek learning. This was the development of Christianity.[23]

Burton gives some alleged details:

> Not content with eradicating "pagan science" by the torch, Christian mobs murdered many of the Museum's scholars in the streets of Alexandria. Such was the fate of the first prominent woman mathematician, Hypatia.[24]

In a similar vein, Ball writes that Hypatia

> was murdered at the instigation of the Christians in 415. The fate of Hypatia may serve to remind us that the Eastern Christians, as soon as they became the dominant party in the state, showed themselves bitterly hostile to all forms of learning.[25]

Neither Burton nor Ball mentions the fact that there were Christian mathematicians at the time — such as Anatolius, bishop of Laodicea, who wrote a book called *Elements of Arithmetic*.[26]

Another example of an anti-religious historian is E. T. Bell. Bell is very unhappy that Blaise Pascal did philosophy of religion:

> we shall consider Pascal primarily as a highly gifted mathematician who let his masochistic proclivities for self-torturing and profitless speculations on the sectarian controversies of his day degrade him to what would now be called a religious neurotic.[27]

Hollingdale follows suit. In *Makers of Mathematics*, we read that Pascal's

> outstanding intellectual powers were exercised mainly on sterile theological speculations occasioned by the sectarian religious controversies of his day.[28]

---

[23] Burton, *The History of Mathematics,* 2nd edn., p. 223.

[24] Ibid., p. 242.

[25] W. W. R. Ball, *A Short Account of the History of Mathematics* (New York: Dover, 1960), p. 111. For a similar view, see V. J. Katz, *A History of Mathematics,* p. 157.

[26] Eusebius, *The History of the Church* (London: Penguin Books, 1965), p. 253.

[27] E. T. Bell, *Men of Mathematics* (New York: Simon & Schuster, 1937), p. 73.

[28] Hollingdale, *Makers of Mathematics,* p. 151.

Incidentally, some present-day philosophers, such as F. Copleston, rank Pascal's 'profitless speculations' among the greatest works in their subject.

Of course, not every historian of mathematics holds the fanatical view that a person who is interested in God is at best wasting their time. D. E. Smith, for example, compares Fibonacci to St. Francis, praising both for 'bringing new light into the souls of men'.[29] Boyer and Merzbach note that the speculations of the Scholastics, such as Thomas Aquinas, led to the Cantorian theory of the infinite.[30] Indeed, one can argue that, far from being a war between Reason and Religion, the history of mathematics is an intellectual journey in which human beings arrive at a deeper understanding of the divine mind, and of the patterns God has instantiated in the physical universe.

In any case, historians of mathematics should consider the possibility that the alleged conflict between Reason and Religion is a myth, and they should acknowledge that if a writer injects a personal, anti-religious bias into his work, it ceases to be objective.

8.19 Advice

What is Good Way to Write a History of Mathematics?

First, the author should become familiar with different branches of mathematics and different chronological periods. Specialists in Newton should not attempt their general history book until they have learned a substantial amount about medieval mathematics. There is a difference between writing serious history and plagiarising from E. T. Bell.

Second, the author should avoid condemning groups of people because they did not further the cause of mathematics. It is true that fifth century Christians did not produce much original mathematics, but this can be stated without patronising, insulting, or pitying Christians. There is more than one good thing in the world, and the historian is playing a silly moral judge if he or she implies that anything bad for mathematics is bad for humanity.

Third, the historian should avoid anachronism. Statements such as

Euclid proved that $2^{n-1}(2^n - 1)$ is perfect

or

Napier used logs to the base e

---

[29] D. E. Smith, *History of Mathematics*, vol. 1 (New York: Dover, 1958), p. 217.
[30] Boyer and Merzbach, *A History of Mathematics*, p. 294.

need qualification. Euclid did not have modern algebraic notation, and Napier had, not the modern notion of 'base *e*', but a curious kinematic definition of logarithms which, with hindsight, can be seen by *us* to be equivalent to a definition of logarithms to the base *e*.

Fourth, the historian should have a sense of humour. The history of mathematics is not a foolish skit, but it is not a stern triumph of Reason either. Great mathematicians, no less than great politicians, are appropriate targets of humour, and the student should learn that great blunders are not confined to politics.

Finally, the historian should pay attention to the chapter headings. The wrong headings can introduce bias and distortion. As an example, let us consider the mathematics done between the years 500 and 1200. Suppose we divide it up into the following chapters:

1. Chinese Mathematics
2. Hindu Mathematics
3. Arab Mathematics

This division suggests that the Chinese get their own special answers when they solve equations. It suggests that mathematics is somehow tied to nationality or religion when, in fact, it is an area of worldwide cooperation. And what about Omar Khayyam? Of course, he gets stuffed into the chapter on 'Arab Mathematics', but he was not an Arab. A more satisfactory way of arranging the same material would be the following:

1. Number Theory from 500 to 1200
2. Algebra from 500 to 1200
3. Geometry from 500 to 1200
4. Khayyam and the Cubic

Chapter 9

Mathematics and Knowledge

In this chapter we presuppose that there is such a thing as mathematical knowledge (and hence mathematical truth), and we investigate what *kind* of knowledge it is, and how it might be possible to have it. We argue that, in fact, there are many different kinds of mathematical knowledge. It is not the case that all mathematical knowledge is infallible and propositional and a priori.

9.1 Infallible Knowledge

Plato insisted that the only kind of knowledge is infallible knowledge.[1] However, many philosophers would allow that a person can know something even if it is possible that he or she be wrong in similar cases. Smith knows that Marilyn is a woman, even though Smith does occasionally mistake a woman for a man. We thus have two kinds of knowledge: infallible and fallible. An example of the first might be God's knowledge that $2 + 3 = 5$. An example of the second might be Smith's knowledge that 211 is prime.

Defining Infallible Knowledge

Is our knowledge of mathematics fallible or infallible? To answer this question, we should first clarify the notion of 'infallible knowledge'. For Plato,

---

[1] Plato, *Republic* 477e.

the infallibility of infallible knowledge implies that a known proposition cannot possibly be false. For Plato,

> Person A infallibly knows that p
> iff A knows that p, and p is a necessary truth.[2]

This will not do, however, since a person can have a fallible knowledge of a necessary truth, or an infallible knowledge of a contingent truth. For example, Smith knows that

$$12345 \times 67890 = 838\ 102\ 050$$

This equation, we may suppose, is a necessary truth, but Smith's knowledge of it is not infallible, because he obtained it by using his calculator, and his calculator occasionally malfunctions. Again, Smith knows that he exists. The fact that he exists is a contingent truth, but Smith's knowledge of it is infallible.

H. Lehman offers an analysis similar to Plato's. For Lehman,

> the concept of fallible knowledge implies that it is possible that a person knows a proposition that P and at the same time it is possible that P is false.[3]

Thus, on Lehman's view, the fact that Smith knows, fallibly, that

$$12345 \times 67890 = 838\ 102\ 050$$

implies that it is possible that this equation is false. But the equation in question is true in every possible universe.

A second analysis of infallible knowledge is presented by K. Lehrer in the *Theory of Knowledge*. According to Lehrer the infallibility of infallible knowledge resides, not in the necessity of the subject matter, but in the essential errorlessness of the knower. For Lehrer,

> Person A infallibly knows that p
> iff it is impossible both that A believe p and p be false.[4]

---

[2] Using logical symbols, we could express this as $IKp \iff (Kp \wedge \Box p)$. Here IK means 'it is infallibly known that'; p is some variable statement; $\iff$ means 'if and only if'; Kp means 'it is known that'; $\wedge$ means 'and'; finally, $\Box$ means 'it is logically necessary that'. For an intuitionist, mathematical truths are contingent on the existence of human minds, so that, strictly speaking, there is no necessity in mathematics, and the right-hand side of Plato's equivalence is always automatically false.

[3] H. Lehman, *Introduction to the Philosophy of Mathematics* (Totowa, NJ: Rowman and Littlefield, 1979), p. 154. In logical symbols, Lehman's analysis reads $Kp \wedge \neg IKp \implies \Diamond(Kp \wedge \Diamond \neg p)$. Here the $\neg$ mean 'it is not the case'; $\implies$ means 'implies'; and $\Diamond$ means 'it is possible'. The consequent of the implication implies $\Diamond\Diamond \neg p$ and hence, in modal logic $S5$, $\Diamond \neg p$.

[4] K. Lehrer, *Theory of Knowledge* (Boulder: Westview Press, 1990), p. 45.

Unfortunately, this definition implies that any person A has infallible knowledge of every necessary truth, since it is impossible that a necessary truth be false, and *a fortiori* impossible both that it be false and A believe it.

Lehrer suggests an improvement along the following lines:

> Person A infallibly knows that *p*
> iff (1) it is impossible both that A believe *p* and *p* be false, and also (2) it is impossible both that A not believe *p* and *p* be true.

This is logically equivalent to

> Person A infallibly knows that *p*
> iff it is a necessary truth that: A believes *p* just in case *p* is true.[5]

This definition, also, can be defeated by counterexamples. Smith infallibly knows that he exists, but, in some possible universes, he is mentally deficient, and incapable of forming any beliefs whatsoever. Hence it is possible that Smith exist but not believe he exists. Again, because God is necessarily omniscient, it is necessary that God believe that $2 + 3 = 16$ just in case $2 + 3 = 16$. For there is no possible universe in which $2 + 3 = 16$ and also no possible universe in which God believes that $2 + 3 = 16$. Hence, according to Lehrer's definition, God infallibly knows that $2 + 3 = 16$.

Perhaps, to define infallible knowing, it would be more illuminating to shift our attention to the *way* of knowing:

> If W is a way of coming to know things,
>
> Person A infallibly knows that *p* thanks to W
> iff
> A believes that *p* because he or she used W,
> and it is necessarily true that if A believes that *p* because he or she used W then A knows that *p*.[6]

Moreover,

> Person A infallibly knows that *p*
> iff for some way W of coming to know things, A infallibly knows that *p* thanks to W.

---

[5] Thus, in logical symbols, Lehrer's improved view is equivalent to $IKp \iff \Box(Bp \Leftrightarrow p)$, where Bp means 'it is believed that'.

[6] In logical symbols, this could be written $IK_W p \iff B_W p \land \Box \forall q (B_W q \Rightarrow Kp)$.

Finding Infallible Knowledge

Granted the 'way of knowing' definition of infallible knowledge, is there any infallible knowledge in mathematics?

Consider Smith's knowledge of the fact that

$$12345 \times 67890 = 838\ 102\ 050$$

There are several ways in which Smith might have come to know this.

(1) He might have read it in a book.
(2) He might have used his calculator.
(3) He might have 'seen it in a flash'.
(4) He might have done some mental arithmetic.
(5) He might have done a multiplication on a bit of paper, and checked his answer by multiplying the numbers in reverse order.

None of these common ways of knowing mathematical facts is infallible. Books contain misprints. Calculators break. Human intuition errs. People make careless mistakes. Pencil marks can be hard to read. And checking, although it vastly decreases the possibility of error, does not eliminate it — since, in mathematics at any rate, two errors can reinforce each other.

What about simpler mathematical truths, such as

$$A \cap B \subseteq A$$

or

Every cube has 8 vertices?

Is 'studying the concepts' a way of having infallible knowledge of basic facts such as these? Not necessarily. Even under normal circumstances, elementary errors are possible. Teachers of set theory know that students sometimes mix up set intersection and set union, and there is no guarantee that a study of the concepts will lead to the correct answer to a test question such as

Circle the theorem among the following 3 statements:

1) $A \cup B \subseteq A$

2) $B \subseteq A \cap B$

3) $A \cap B \subseteq A$

Note also that 'studying the concepts' led mathematicians to false conclusions such as

There is no everywhere continuous nowhere differentiable function

or

for any property, there is a set with just the things having that property as its elements.

In conclusion, it seems that the sceptics have to be granted that we have no way of acquiring infallible knowledge of typical mathematical facts. And the view that mathematical knowledge is certain is wrong. Indeed, as we noted in Chapter 8, the history of mathematics is filled with blunders, even about the most basic matters!

Self-Evident Truths

Granted that our knowledge of mathematics is not typically infallible, is it still possible that at least *some* mathematical knowledge is infallible? For example, what about simple 'self-evident' statements such as

$$2 = 2$$

or

all squares are quadrilaterals?

Surely, anyone who understands such a statement is justified in believing it (just because he or she understands it), and, surely, believing it on the basis of understanding it is a way of knowing it, and knowing it infallibly.

Note, however, that it is not always self-evident whether or not a statement is self-evident.

Consider Euclid's Parallel Postulate. This, recall, is the assertion that if two straight lines are tilted towards each other, they will cross each other, assuming they are extended sufficiently far. Is the Parallel Postulate a self-evident feature of straightness? Or is it something that ought to be proved from more basic assertions? Even before Beltrami proved the consistency of hyperbolic geometry, mathematicians were uncertain about how self-evident the Parallel Postulate was, and, even today, one can argue that the Parallel Postulate is a defining mark of *real* straightness, as opposed to the hyperbolic substitute.

Another example is Frege's principle that,

for any property, there is a set with just the things having that property as its elements.

Before the year 1900, this seemed self-evident, but, far from being an item of infallible knowledge, it led to the contradiction called the Russell Paradox.

In conclusion, although self-evidence does give rise to infallible knowledge (in certain simple cases), it is not unproblematic.

In his book *Understanding the Infinite,* Lavine makes the surprising claim that, at least relative to a view of sets according to which they are well-ordered, the Axiom of Choice is self-evident.[7] This claim is surprising because the Axiom of Choice involves the infinite, and, given all the paradoxes connected with the infinite, one might argue that, as far as the infinite is concerned, there is nothing self-evident. Lavine supports his claim by showing how one might arrive at the Axiom of Choice from a merely finite version of it, using 'extrapolation'. However, it is not clear whether Lavine's extrapolation gets one to the infinite — as opposed to the indefinitely large but still finite.

Renouncing Infallible Knowledge

People can have satisfying and fruitful relationships with other people, even though they do not have an infallible knowledge of the character or habits of those other people. Similarly, people can have satisfying and fruitful relationships with mathematical objects, even though they do not have absolute certainty about the nature or properties of those objects. Infallibility is like other kinds of perfection: human beings can live quite well without it. This is not to say that we never have infallible knowledge about anything — Smith does have infallible knowledge of some basic facts, such as the fact that $2 = 2$ — but it is to say that our lack of infallible knowledge is not important. We need health, we need love, and we need information — but we can do quite well without infallibility.

9.2 Knowledge by Acquaintance

In *Mysticism and Logic,* Bertrand Russell describes knowledge by acquaintance as a direct awareness of an object, or a memory of such an awareness.[8] For Russell, we have knowledge by acquaintance of sense data, and also of universals (e.g. before-and-after, triangularity), but not of physical objects, or other minds. Russell does not say whether we can have knowledge by acquaintance of sets, numbers, or geometrical configurations.

There are three possibilities. (1) We have knowledge by acquaintance of certain mathematical objects. (2) We have knowledge by acquaintance of

---

[7]Lavine, *Understanding the Infinite,* pp. 110, 116, 160, 242, 303.

[8]B. Russell, *Mysticism and Logic* (London: Longmans, Green and Co., 1925), ch. X.

something that bears to mathematical objects a relation analogous to the relation between sense data and physical objects. (3) We have no knowledge by acquaintance of anything mathematical.

Since it is impossible to be acquainted with something that does not exist, possibilities (1) and (2) are incompatible with mathematical nihilism. Possibility (3), on the other hand, is incompatible with certain statements, often made by mathematicians, such as

> I am acquainted with such configurations

or

> I recognise a group when I see one.

Opposed to knowledge by acquaintance is propositional knowledge, that is, knowledge that a particular statement is true. Beautiful Bina Ngan knows, for example, that the ruler of Russia is powerful, even though she has never made his acquaintance.

Those who view mathematics as an exploration of a Platonic heaven may be drawn to the position that some mathematical knowledge is knowledge by acquaintance, but those who understand mathematics as a collection of axiomatic systems built around undefined terms may want to hold that all mathematical knowledge is propositional.

In the *Theaetetus,* Plato suggests that propositional knowledge be characterised as 'true belief with the addition of an account'.[9] Today, we might put it this way:

> Person A knows that p
> iff
> p, and A believes that p, and A is justified in believing that p.

Although the above conditions are necessary for propositional knowledge, E. Gettier has shown that they are not sufficient.[10] For example, suppose that, as a hoax, a well-respected newspaper prints a story about the President of the University killing her lover. Suppose that, unknown to the editors of the newspaper, the President has, in fact, just killed her lover. Then a reader of this well-respected newspaper has a justified true belief that the President has killed her lover. Yet, the reader does not *know* that this has occurred, because his belief is based on a hoax.

To obtain conditions which are sufficient, as well as necessary, for propositional knowledge, philosophers have examined many possibilities. Lehrer, for example, suggests a formulation along the following lines:

---

[9] Plato, *Theaetetus* 201d.
[10] E. Gettier, 'Is Justified True Belief Knowledge?', *Analysis,* 23 (1963), 121-3.

Person A knows that p
iff
p, and A believes that p, and A is justified in believing that p,
and A's justification does not rest, in any essential way, on some
false belief A has.[11]

In terms of this definition, the newspaper reader does *not* know that the
President has committed murder, because the reader's justification for this
correct belief rests on his false belief that the newspaper is not publishing
a hoax.

One could engage in a lengthy discussion about the merits of such a
definition. For example, one could investigate whether a person A really
knows that p if, although the above four conditions are met, A also has a
false belief that his justification for believing that p rests, in an essential
way, on a false belief. We shall not pursue such possibilities here, but,
instead, turn to the question, 'to what extent is mathematical knowledge
knowledge by acquaintance?'

Acquaintance with Mathematical Objects

Much mathematical knowledge is propositional. Professor Lambek's knowledge that there are *at most* five regular solids is a justified true belief whose
justification does not depend on any false belief he has. However, what
about his knowledge that there are *at least* five regular solids? The belief
that there are at least five regular solids can be justified by a rigorous,
nonintuitive proof, but it can also be based on an acquaintance with the
cube, tetrahedron, octahedron, icosahedron, and dodecahedron.

Many mathematical theorems are known propositionally because the
reality they describe is known by acquaintance, and this acquaintance is
what justifies the true belief. Consider the following 'fixed point' theorem:

If f is a continuous function on the interval $[0, 1]$, with $f(0) = 1$
and $f(1) = 0$, then there is at least one number x between 0
and 1 such that $f(x) = x$.

The proof is in the picture. To get from $(0, f(0))$ to $(1, f(1))$, the graph
of f has to cross the line $y = x$, and, where it does, we get a 'fixed point'
(x, x). Our acquaintance with the topology of the figure justifies our belief
in the theorem. We know *that* the theorem is true, because we *'see'* the
curve $y = f(x)$ crossing the straight line $y = x$.

If we consider mathematics from the 'axiomatic' standpoint, it may at
first seem that any knowledge by acquaintance is irrelevant. It may at first

---

[11]Lehrer, *Theory of Knowledge*, p. 138.

seem that the only thing that matters is a knowledge of implications. However, mathematicians do not normally choose axioms in a purely arbitrary manner. They pick them so that they can deduce certain intuitively known truths from them.

Suppose, for example, that a mathematician is trying to organise the insights of Euclidean plane geometry. One of these insights is expressed by the theorem that the square on the diagonal of a given square has twice the area of the given square. As Plato suggests in the *Meno,* this insight is part of a knowledge by aquaintance of a geometric figure — one consisting of a square whose sides' midpoints are joined in all possible ways, dividing the square into 8 right triangles. Now if the mathematician finds that his axioms do not let him deduce the theorem that expresses this knowledge by aquaintance, he will simply revise his axioms until they do. The propositional knowledge embodied by the axioms is typically dependent on some knowledge by acquaintance of concepts or configurations.

Of course, a mathematical nihilist will object to this way of putting it. For the nihilist, the mathematician is subjugating his axioms, not to any acquaintance with mathematical objects, but merely to some preferred fantasy. The mathematician talks about 'insights' or 'truths', but the nihilist dismisses this as so much prejudice and illusion.

Knowing How

There is also 'knowing how'. 'Knowing how' is an ability to accomplish something without necessarily being able to explain how one does it.[12] Smith knows how to make shortcake, even though he becomes confused when he is asked questions about how he does it, and thereby displays a lack of propositional knowledge. Difficult Douglas knows how to ride a bicycle, but this knowledge is not merely his being acquainted with the bicycle, nor is it his knowing facts about momentum.

There are examples of 'knowing how' in mathematics. The mathematician who is good at, say, finding solutions to Diophantine equations, but poor at explaining his methods, has the 'know how', but not the propositional knowledge.

In a draft of his book *The Conceptual Foundations of Mathematics,* J. R. Lucas points out that knowing how is less vulnerable to sceptical attack than propositional knowledge is:

> Once we see the mathematician as an active operator who does things he knows how to do rather than a passive percipient of eternal truths, mathematical knowledge appears much less

---

[12] G. Ryle, *The Concept of Mind* (London: Hutchinson, 1949), ch. 2.

puzzling. Whereas claims to know that something is the case invite questions "How do you know", which we are hard put to answer, claims to know how to do something are vindicated by actually doing it.

9.3 A Priori Knowledge

A priori knowledge is knowledge which does not depend, in any essential way, on sense data. A posteriori knowledge is knowledge which does depend on sense data. For example, consider the statement

> There are at least three ways of tiling the plane using regular polygons.

I may know this because a mathematician told me, or because I have seen pictures of the triangular, square, and hexagonal tilings. However, perhaps I could have learnt this truth without the help of my five senses. Perhaps I know it because I once worked it out for myself that it is a logical consequence of Hilbert's abstract, axiomatic definition of the plane. If I have, or could have, acquired my knowledge in this latter way, then it is, at least in principle, independent of sense data, and hence a priori.

In order to grasp the concept of a priori knowledge, it is useful to distinguish it from infallible knowledge. First, 'a priori' does not imply 'infallible'. For suppose I mentally multiply 41 by 11 to obtain 451. Then I know a priori that

$$41 \times 11 = 451$$

However, I do not know this infallibly, since I occasionally make mistakes in mental arithmetic. Second, 'infallible' does not imply 'a priori'. For suppose that Professor Smith is stepping on my toe. I know infallibly that someone feels pain, but my knowledge is not a priori, since it depends on the sense data arriving from my foot.

It is also useful to distinguish a priori knowledge from knowledge of necessary truths. Saul Kripke does this in 'Naming and Necessity'.[13] He notes that one can have a priori knowledge of facts that are not logically necessary. For example, one can know a priori that

> The standard metre bar is one metre long

---

[13] S. Kripke, *Naming and Necessity* (Cambridge: Harvard University Press, 1980).

— although it is not a necessary truth that this is so. Indeed, it is not even necessary that there be a standard metre bar.[14] On the other hand, as Kripke also points out, the statement

Hesperus is just Phosphorus

is a necessarily true identity statement (about the planet Venus), even though there is no way of knowing this that is not based on sense data.

Some thinkers (often of an empiricist persuasion) claim that there is no a priori knowledge whatsoever. One argument goes as follows:

A person cannot know anything unless he knows a language.
No one can know a language unless he receives sense data.
Thus all knowledge is dependent on sense data.

One way to block this argument is to qualify the meaning of a priori knowledge as follows:

A priori knowledge is knowledge which, at least in principle, can be acquired independently of sense data — with the possible exception of sense data necessary to acquiring a language.

For example, consider the statement

None of my married friends is a bachelor

or the statement

$x = y$ and $y = z$ implies $x = z$.

Knowledge of these truths may presuppose a certain a posteriori linguistic knowledge, or a certain a posteriori ability to understand and relate concepts, but this does not count against their a prioricity.

Is Mathematical Knowledge A Priori?

So far the only examples of a priori truths we have given are 'definitional' or 'analytic' in some way. What philosophers usually want to know when they ask about a priori knowledge in mathematics is whether there are any more substantial examples. It seems there are. For example, consider the statement:

---

[14] Other examples of a priori knowledge of contingent truths would be (1) my knowledge that I exist, and (2) God's knowledge that there are supercompact large cardinals (assuming there are some).

> There are infinitely many sets.

This does not follow from the definition of 'infinite' or 'set', and yet it is usually taken to be something we know in mathematics. Let us assume that this is so. Now is this knowledge a priori without being definitional?

An empiricist might reply that it is not a priori because it is based on our knowledge that the usual set theory is an essential component in the only scientific theories which (almost) always give correct predictions about the natural world.

A logicist might reply that it is definitional because it is an axiom of set theory, and the axioms of set theory do no more or less than define words like 'set'.

In regard to the empiricist reply, an a priorist could say that scientific theories have nothing to do with it. The reason one knows there are infinitely many sets is that one knows that the usual set theory is an essential component in any explanation or account of mathematics. Or the reason one knows there are infinitely many sets is that one has a reliable intuition to the effect that every set is a proper subset of some other, 'larger' set.

In regard to the logicist reply, an a priorist could say that there is a difference between defining a set in such a way that any two sets have an intersection, and defining a set in such a way that, say, the Continuum Hypothesis is true. The second 'definition' goes far beyond basic logic. Call it 'definitional' if you will, but our knowledge that there are infinitely many sets is a nontrivial example of a priori knowledge in mathematics.

A Posteriori Mathematical Knowledge

To some it may seem that, although human beings typically acquire mathematical knowledge using sense data, this sense data is not ever *essential* to the acquisition of that knowledge, and hence *all* mathematical knowledge is a priori. After all, God did not learn something new about mathematics by deciding to create the universe or by deciding to become incarnate.

This is a nice other-worldly view, but there is a counterexample to it. A rational being who lacked any access to sense data would not have the experience of spatiality that is presupposed in a complete knowledge of, say, continuity, direction, or dimension. This rational being might, of course, understand an arithmetical model of geometry, but he (or she) would lack a full comprehension of geometric concepts as geometric. It is one thing to know all about the number 120, including the fact that it is the distance (in kilometres) from Tiberias to Jerusalem. It is another thing to know what it is like to trudge from one hot dusty village to another, always heading in the same boring direction. Again, it is one thing to understand the infinite series

$$z - \frac{z^3}{3!} + \frac{z^5}{5!} - \frac{z^7}{7!} + \cdots$$

— this as an object in abstract analysis — and quite another thing to know it as a ratio of two sides of a mountain, or as a beautiful, sensual curve.

9.4 Knowledge à la Plantinga

According to A. Plantinga, a person A knows that p just in case p is true, A believes that p, and that belief was

> produced by cognitive faculties functioning properly (subject to no malfunctioning) in a cognitive environment congenial for those faculties, according to a design plan successfully aimed at truth.[15]

Plantinga does not say how we might know that a given cognitive faculty enjoys the pleasant properties listed above, and he does not say whether the enjoyment of those properties is a very common occurence. It is clear for example, that the ivory tower is a cognitive environment not congenial for cognitive faculties, but it is not clear whether, say, the slums of Detroit are much better.

One problem with Plantinga's characterisation of propositional knowledge is its neglect of the fact that people who know things are, to some extent, in touch with the realities related to the things they know. For example, suppose that Smith is just a brain in a vat, and suppose all of Smith's 'perceptions' are artificially induced by a team of scientists who, as it happens, work according to a design plan successfully aimed at inducing true beliefs, and only true beliefs, in Smith. Smith himself is unaware of all this. Now one day the scientists make Smith 'experience' sense data of the sort one would experience on Mount Washington — the way it 'really is' — and he forms the true belief that it is cool on the summit even in June. The scientists have not yet got around to inducing such true beliefs as 'the summit is above the tree line' or 'Mount Washington is in North America' — Smith's conception of Mount Washington is still woefully incomplete — but at least he has a true belief about the temperature.

According to Plantinga, Smith *knows* that the summit of Mount Washington is cool in June.

However, consider the following facts. (1) Smith's true belief is based on his false belief that he has actually been on Mount Washington. (2) Smith

---

[15] A. Plantinga, *Warrant and Proper Function* (New York: Oxford University Press, 1993), pp. viii-ix.

is totally at the mercy of the scientists. (3) Smith has no way of checking the information he receives (no way of comparing his 'virtual reality' to the real world). (4) Smith, even if he could talk, would not be able to give you a good reason why you should agree with him about the temperature on Mount Washington: he would merely say something false like 'I was up there yesterday, and it was cold'. (5) Smith's belief about the temperature is undermined by his ignorance of facts one normally knows when one knows the temperature. 'Smith, you say you went up this mountain and it was chilly, but you do not even know if the mountain is in North America!'

Given the above facts, I think one should not ascribe knowledge to Smith. If a person is in a vat (or an ivory tower) — and he does not even realise he is in a vat (or an ivory tower) — then he is too out of touch with the external world to have knowledge of it. The information he receives is like a package delivered by Canada Post: it arrives in a distorted shape, and parts are missing, and you remove the remaining wrapping, but you still do not know what it was that was sent to you. No doubt the clever scientists can arrange matters so that, by sending you original article A (e.g. a television), you will actually get the article they intend you to have, namely, B (e.g. some broken glass) — but you yourself are best described as being 'in the dark'.

Let us consider a case in which Plantinga discusses knowledge of mathematics. Suppose Marilyn's cognitive system has been enhanced so that she can tell at once whether a natural number less than, say, one million is prime. When she thinks about 67, it seems obvious to her that 67 is prime — just as when we think about 6, it seems obvious to us that it is even. This enhancement has been carried out according to all the Plantinga requirements, so that, for example, the resulting system is 'successfully aimed at truth', and, indeed, Marilyn never mistakes a prime for a composite or vice versa. According to Plantinga, if this situation obtains, then Marilyn knows that 67 is prime.[16]

Unfortunately for Plantinga, nothing has been said about whether Marilyn has any *contact* with what she allegedly knows. Suppose we take Marilyn, put her into one of those ideal cognitive environments (assuming there are any available) and ask her 'how do you know 67 is prime?' She replies

I just see it has no factors.

'And how do you see that?' we ask.

Well, it ends in 7, so 2 does not divide into it, so it's prime.

[16] A. Plantinga, *Warrant: the Current Debate* (New York: Oxford University Press, 1993), p. 59.

'Couldn't 3 divide into it?'

> Let me think. No, 3 would leave a remainder of 2. So it must
> be prime.

'Well, could 5 go into it?'

> No, because, 5 does not go into any number beginning with 6.
> There, we've covered 2, 3, and 5. There is no need to ask me
> any more questions. 67 is prime and that's all there is to it.

'That's not all there is to it. To know that 67 is a prime, you need a warrant
for your knowledge. You need at least to be able to say something like 'I
know it's true, because I got it from my enhanced system, and my enhanced
system is reliable'.

> What enhanced system? I don't know what you are talking
> about.

There is, of course, a warrant for Marilyn's beliefs which is implicit
in the 'proper functioning' of the 'properly functioning' epistemic system
etc., but nothing in Plantinga's characterisation of knowledge implies that
Marilyn have any access to that warrant. Plantinga's alleged knower can
technically 'possess' a 'warrant' for the alleged knowledge, but be totally
out of touch with what is going on.

One is reminded of a computer, a computer that has been given a list
of all the primes less than a million and programmed so that it can tell
you whether any given number is on that list. However, if you ask it any
other question, the only response it can make is to print out the definition
of 'prime'. It cannot even give you a reference to a number theory book.

Plantinga has another example which illustrates the fact that, if he is
right, a knower can be totally out of touch with what he or she knows.
Plantinga imagines a planet whose inhabitants are so designed that, whenever they see a church-bell, they see it as orange.[17] Happily, conditions
on this planet are such that, indeed, all church-bells are orange, and the
epistemic system of these aliens is, in fact, successfully aimed at the truth
(within that peculiar congenial environment) — and meets the other requirements Plantinga has in mind. None the less, if one of the distinguished
campanologists on that planet were to visit earth, he would soon become
known for his ignorance. Great Tenured Professor of Campanology he may
be, but he is out of touch with reality. A similar thing must be said of
Plantinga.

---

[17]Plantinga, *Warrant: the Current Debate*, pp. 62-63.

In the history of mathematics, there have been people, like Ramanujan, who have seen deep and difficult things 'in a flash', and have had trouble explaining their insights. It was, a little, as if their magnificently functioning epistemic systems were somewhat beyond even them. However, that is not to say that they were out of touch with their epistemic systems, or out of touch with their subject matter. Ramanujan did prove some of his theorems, he checked all the others numerically, and he was able always to relate his insights to other insights (one of which might, say, be deduced from the other). He had an intuitive contact with mathematical objects, and it was this contact, more than anything else, that provided the foundation for his knowledge. Far from being 'out of touch' with his theorems, Ramanujan had a unbreakable grasp of them. Indeed, Ramanujan 'knew' his numbers in a sense analogous to that in which Abraham 'knew' his wives: he was 'intimate' with them.

On Plantinga's view, we could hook up a dog's brain to a pocket calculater in such a way that the properly enhanced dog would form true beliefs, and only true beliefs, about numbers being prime. This dog could bark out its recognition of primes, taking only 6 seconds to recognise the primality of a prime with 6 digits. It could win prime recognition speed contests against eminent mathematicians (such as Ramanujan), and, best of all (according to Plantinga) it would know that what it was barking about was prime numbers. It might not be very good at division, it might have no idea that it had been enhanced, but it would know that 67 is prime.

One has to credit Plantinga for the imagination that goes into his thought experiments. I believe it is his own original idea that there might be creatures possessing knowledge in spite of being totally out of touch with the objects of their knowledge and also totally out of touch with their own (ever so properly functioning) epistemic systems. One hopes that Plantinga will turn from philosophy to fiction.

9.5 Computer Knowledge

Much mathematical knowledge is acquired by, and based on, not science fiction mind enhancement, but reasoning — so, in order to get back to epistemology, we shall now say something about reasoning.

There are two ways of understanding reasoning, and two ways of characterising the corresponding forms of knowledge. On one view, reasoning starts with some known 'axioms' and proceeds as rigorously as possible, in small mechanical steps, to the conclusion. A person who 'follows the reasoning' may not have the slightest idea how it was created, or what ideas it is expressing, but they will none the less come to a kind of knowledge of the conclusion. Indeed, there is no particular need for them to 'follow the reasoning' to do so. It is even better if they get the computer to go through the small mechanical steps, since the computer will more reliably check the validity of each step, and then they can take as known whatever the computer concludes. This happens in the case of some highly computational proofs in graph theory and number theory. We 'know' that the Four Colour Theorem is true, or that there are more than 30 perfect numbers, because the computer has verified a sequence of steps too long for any one human to follow. Let us call knowledge obtained in this way 'computer knowledge'.

On a second view of reasoning, what is important is awareness and insight. The deductive steps may be quite large if a steady intuition allows the mind to travel from one idea to the next. The reasoner gets to do modus ponens, of course, but he or she also gets to 'weigh' reasons, to choose which ones are important, and which ones should be ignored for the time being. The reasoner is not bound to precisely defined first order procedures, but can use any mixture of higher logic, meta-logic, and even metaphysics. The emphasis is not on a sequence of well formed formulas, but on a mind seeking enlightenment.[18] It is this second sort of reasoning one finds in a research seminar in, say, number theory. The steps are usually just sketched, while it is the ideas that are discussed and debated. Everyone is striving to share intuitions, to 'see' what is going on. You hear things like:

> I don't know the exact steps, but I am morally certain one could prove this using some standard homological algebra

or

> the function does not have much place to go, and if the set is still too large, we could get the result by modding out by the various relations.

This is not to deny that the mathematicians then go home and try to work out some of the minute details — possibly using a computer for this task — but it does mean that working mathematicians do not reason the same way a computer does. Even at home, the mathematician sees no particular need to work out *all* the details — just a few selected details which he or she considers especially interesting or problematic. Anyway, if we have to name things, we could give the label 'humanist knowledge' to the sort of knowledge obtained through the reasoning typically used by working mathematicians.

---

[18] See W. S. Anglin, *Free Will and the Christian Faith*, p. 11.

9.6 Conclusion

We know mathematical truths in a variety of ways. Many different sorts of knowledge are represented in mathematics. There is Smith's infallible knowledge that $2 = 2$, and his fallible knowledge that

$$12345 \times 67890 = 838\ 102\ 050$$

There is Smith's knowledge by acquaintance of five regular solids, and his propositional knowledge that there are no other regular solids. There is Smith's a priori knowledge that there are infinitely many counting numbers, and Smith's a posteriori knowledge that a circle has an interior and an exterior. Mathematical knowledge is restricted to things mathematical, but it is not restricted in other ways.

9.7 Appendix: Terms in Epistemology

Since different philosophers use epistemological words in different ways, it might be useful to include the following glossary, indicating how I use some of those words.

acquaintance = familiarity; knowledge by acquaintance is knowledge of an object (possibly a state of affairs) by means of a perception or intuition of it

actual world = reality; the possible world which happens to be actual; a merely possible world is a possible world such that at least one false proposition is true in it; the actual world is the possible world such that every proposition true in it is true

a posteriori = relying on the five senses; Diana has a posteriori knowledge that Fermat's Last Theorem is true, since she takes it merely on the word of a famous mathematician

a priori = without using sense data in an essential way; Professor Richard Taylor has a priori knowledge of the truth of Fermat's Last Theorem

contingently true = true but not necessarily true; one usually knows a proposition that is contingently true empirically, a posteriori

empirical = restricted to the five senses; without empirical knowledge, one cannot do much chemistry, but one can do a lot of logic

fact = an actual state of affairs; a component of the actual world

impossible world = a universe for which there is a proposition p such that, if that world were actual, both p and not-p would be true

intuit = perceive, not with one of the five senses, but with the mind or soul

judgement = decision to believe a proposition

justification = reason or evidence for believing a proposition, and, in some cases, entitling one to claim one knows that proposition

meaning of a linguistic entity = information it conveys, either on account of its linguistic components, their semantics, and the grammar of their composition, or on account of some other way the speaker uses it to communicate what he or she has in mind

necessarily true = true in all (logically) possible worlds; there is nothing surprising about a priori knowledge of necessary truths

necessary object = one such that it is necessarily true that that object exist (e.g. the number 2)

object = entity; could be abstract or concrete; could be a structure or a person

possible = true in at least one possible world

possible world = state of affairs such that, for any proposition p, exactly one of p and not-p is a correct description of that state of affairs

proof = deduction from premisses; one can know the premisses (e.g. axioms of arithmetic) without necessarily knowing the conclusion of the proof (e.g. prime number theorem); anything can be 'proved' — it is just that the proof may be unconvincing, circular, based on false premisses, etc.

proposition = description of state of affairs; equivalence class of declarative sentences with same meaning; that meaning itself; something to which one can literally ascribe truth or falsehood; a true proposition describes a fact; propositional knowledge is knowledge of facts; a proposition is timelessly true or timelessly false; 'Bill is coming' is a sentence which expresses a proposition such as: at 3 PM, Bill is getting closer to point P

self-evident = true, and seen to be true just by understanding it

sense = meaning

sentence (declarative) = a linguistic entity which is related to a proposition as a token to a type

situation = state of affairs

state of affairs = situation; referent of a declarative sentence; component of a possible, or even impossible, 'world' (i.e. universe); states of affairs are not linguistic entities; states of affairs are abstract objects, some actual, some merely possible, some impossible; they exist objectively; two states of affairs are incompatible if the propositions describing them are not simultaneously true in any possible world; one can have propositional knowledge that Diana is divorced, and also knowledge by acquaintance of Diana's being divorced (one is 'aware of the situation'); a state of affairs obtains timelessly; 'the state of affairs in Germany' is short for 'the state of affairs in Germany at time t' (for some t)

statement = proposition

true = corresponding to reality; if someone knows that p then p is true

true in a possible world = would be true if that world were the actual world

warrant = name given by Plantinga to whatever it is that turns true belief into knowledge; justification

world = universe; could be actual, merely possible, necessary, or impossible

On my view, 'Bill is coming' is a sentence expressing a proposition like

> at 3 PM Bill is getting closer to point P

and this proposition describes a state of affairs consisting of Bill's getting closer to that point at that time, and this state of affairs picks out some features in a world, a world in which Bill is walking towards P — the location of the speaker of the sentence — and this at 3PM, etc.

Chapter 10

Mathematics and Education

One of the very few things most philosophers agree about is that mathematics is an essential part of a good education. In this chapter we look at several philosophies of education, examining their implications for the teaching and learning of mathematics. We also consider the importance of Euclidean geometry in education.

10.1 Education for the Economic Status Quo

On one view, the goal of education is to maintain the economic status quo. On this view, there are two sorts of education, one for future workers, and another for future managers. The educational system has to separate these two groups, and give them the particular knowledge they will need. Mathematics plays a key role in the separating process, because the ability to score well on Math tests is a reliable indicator of managerial talent.

The educational system, on this view, must train the future workers to accept the authority of the future managers. It must train the future workers to accept tasks which are boring. Here, again, mathematics can play a key role, if it is taught in such a way that the future workers are trained to 'follow formulas', without bothering themselves about why the formulas are true. Future workers need basic counting skills, and sufficient skills in estimation to know when their calculators are broken, but what they most need is drill in the rote application of formulas. This teaches them to follow orders without asking deep questions. Mathematics courses for future workers should be structured so that there is little time or incentive to have a clear understanding of the material. It should be common knowledge that all the test questions are straightforward applications of formulas — or questions involving the regurgitation of memorised proofs.

Mathematics can also play a key role in the training of managers. Taught as an exploration of reason, mathematics can stimulate the imagination, leading it to envisage new ways of organising the world. A sound knowledge of business mathematics will help future managers understand various economic situations, and aid them in developing ideas for changing those situations. In teaching mathematics to future managers, it is better to concentrate on just a few formulas, spending the time to ensure that the underlying ideas and connections are fully grasped. The teacher should have the students derive, for themselves, a number of different proofs for key results. No time should be wasted on memory work.

In *The Philosophy of Mathematics Education,* Paul Ernest discusses education for the economic status quo under the headings 'industrial trainer ideology' and 'technological pragmatist ideology'. Ernest gives several arguments against these views, one of which is that they are self-defeating. Gearing education to maintain the economic status quo is not an effective way to maintain the economic status quo. This, says Ernest, is because industry is changing in such a way that it will soon need more and more workers who can bring insight and initiative to their jobs.[1]

10.2 Education for Democracy

On this view, the aims of education should be subservient to the aims of democratic government. Students should be taught the kind of mathematics that will help them make intelligent decisions about national policies. A knowledge of statistics, game theory, and actuarial science would be useful. Moreover, the way the mathematics is taught should be consistent with the development of independent-minded but community-oriented citizens. There must be no bias against any gender, race or class.

Ernest discusses a version of this view under the heading 'public educator ideology'. According to Ernest,

> This theory sees children as needing to actively engage with
> [sic] mathematics, posing as well as solving problems, discussing
> the mathematics embedded in their own lives and environments

---

[1] P. Ernest, *The Philosophy of Mathematics Education* (London: The Falmer Press, 1991), pp. 151 and 165.

(ethnomathematics) as well as broader social contexts. Learner (and teacher) conceptions and assumptions need to be articulated, confronted with other perspectives, and challenged, to allow the development of critical thinking. This leads to conflict, which is necessary for the accommodation and growth of new conceptions.[2]

By 'conflict', Ernest means conflict over what the theorems of mathematics are (he is a fallibilist), conflict over the importance of different kinds of mathematics (girl-friendly mathematics versus male dominated mathematics), and conflict over the report card (with possible negociation in cases of racially based dispute). Ernest notes that, although many people (at least in his own country) will agree that conflict is necessary in democratic politics, not many people will agree with the idea of taking the basic algebra course for teenaged schoolchildren and transforming it into a series of political confrontations. Note, however, that this is not a problem with education for democracy per se, but with Ernest's somewhat extreme version of that philosophy.

10.3 Education for the Development of the Individual

A more person-oriented view of education is the view that it is a process whereby each individual reaches their 'full potential'. This 'full potential' can mean a number of things. It can mean awareness of one's place in history or general culture. It can mean a psychological integration, such as that achieved through Jungian psychoanalysis. It can mean a spiritual union with God, in which one 'sees' the Truth himself. On many interpretations, one's 'full potential' involves the ability to reason, and to understand the reasoning of great thinkers. Here a knowledge of mathematics is useful, if not essential.

10.4 Plato's View

In Plato's ideal political state, there are three classes: workers, soldiers, and rulers. The only people who study a significant quantity of mathematics are those being trained to be rulers, and for them it is absolutely essential. They study it, in depth, for ten years.

---

[2]Ernest, *The Philosophy of Mathematics Education*, p. 208.

The rulers must know mathematics because it is useful in warfare and administration.[3] They must know mathematics because, taught the right way, it provokes thought, raises questions, and stimulates reasoning.[4] Finally, they must know mathematics since it is this knowledge alone which can begin the process to 'convert the soul to the contemplation of true being'.[5] By 'true being', Plato means the things that are truly important, such as the Form of Goodness. Plato wants the rulers to be highly developed individuals, so that they will be the best possible leaders for the state.[6] This they will be only if they acquire a knowledge of the Good, but, to acquire this knowledge, they need to do some proper philosophy, and for proper philosophy the prerequisite is mathematics.

Although not everyone would endorse Plato's ideas in their entirety, most would agree that a government should be run by people who are clear-headed enough to do mathematics. Government, moreover, requires planning, and planning requires an ability to interpret statistical data, and an ability to estimate, mathematically, the results of various possible decisions.

10.5 John of the Cross's View

For John of the Cross (1542–1591), developing one's full potential is a spiritual transformation, the goal of which is intimacy with God. As far as this transformation is concerned, intellectual knowledge, such as mathematics, is, at a certain stage, harmful. Rather than raising the soul to God, it ties the soul to a lesser good. In the *Ascent of Mount Carmel*, John of the Cross affirms:

> Anyone, therefore, who values his knowledge and ability as a means of reaching union with God is highly ignorant.

However, anyone fortunate enough to 'depart from intellectual reasonings, walks securely'.[7] John of the Cross adds that faith

> deprives and blinds a person of any other knowledge or science by which he may judge it. Other knowledge is acquired by the light of the intellect, but not the knowledge that faith gives.

---

[3] Plato, *Republic* 522e, 526d.
[4] Ibid., 524d-525a, 526b.
[5] Ibid., 525a, 527b.
[6] Ibid., 519c-520e, 540a-b.
[7] John of the Cross, *The Collected Works of St. John of the Cross*, trans. K. Kavanaugh, and O. Rodriguez (Washington: ICS Publications, 1979), pp. 79 and 108.

Faith nullifies the light of the intellect, and if this light is not darkened, the knowledge of faith is lost.[8]

Although John of the Cross did not rule out the possibility that, once a person had achieved intimacy with God, they might then know a great deal of mathematics, he does maintain that a knowledge of mathematics can impede the achievement of this intimacy. For education in the sense of developing one's 'full potential', one should, at a certain stage, renounce mathematics.

10.6 Comte's View

Auguste Comte (1798–1857) took a 'scientific' view of human beings, according to which we are completely subject to the natural laws governing physical objects. In order to understand ourselves, as well as the universe generally, we should attempt to discover these laws, and, since these laws are mathematical, we must learn mathematics. In his *Cours de Philosophie Positive,* Comte writes:

> Thus it is by the study of mathematics, and only by this study, that one can form a correct and deep idea of what it is to be a science. It is only there that one should seek to know with precision the general method that the human spirit constantly uses in all its positivistic research, because there is nowhere else that questions are solved in such a complete manner, and deductions continued to such an extent with severe rigour. It is there, also, that our understanding has given the greatest proofs of its power.[9]

Given Comte's glorification of science, it is not surprising that he makes mathematics the foundation of his ideal curriculum. Comte puts particular emphasis on calculus, moreover, since calculus is useful in physics.

10.7 Smith's View

D. E. Smith gives several reasons for studying mathematics. For one thing, it has practical applications; for another, it sharpens the wits.[10] A knowl-

---

[8] Ibid., p. 111.

[9] A. Comte, *Cours de Philosophie Positive,* vol. I (Paris: Anthropos, 1968), p. 108. My translation.

[10] D. E. Smith, *The Teaching of Elementary Mathematics* (New York: Macmillan, 1907), pp. 1 and 15.

edge of mathematics also helps one understand 'ethical, religious, and philosophical thought'.[11]

Smith is intrigued by the idea that mathematics should be learnt in the order in which, historically, it was discovered. This would be more natural, easier, and more efficient. Also the student would acquire an insight into the history of mathematics, and hence a better understanding of culture.[12] If this idea were implemented, the student would learn unit fractions before decimals, and Euclidean geometry before set theory.

Unlike some other thinkers, Smith holds that knowledge of mathematics is good just for its own sake. He writes:

> The real reason for learning, as it is for teaching algebra, is, that it is part of Truth, the knowledge of which is its own reward.[13]

Smith adds:

> algebra has its ethical value, as has every subject whose aim is the search for truth.[14]

Again, the truths of Euclidean geometry 'enrich us by our mere contemplation of them.'[15]

10.8 Jung's View

According to C. G. Jung, part of developing one's 'full potential' is undergoing the 'individuation process'. This is the process of psychological growth whereby the various components of the soul are united into a harmonious whole. The person becomes aware of what he or she is really like, and learns to handle weaknesses and strengths, achieving a balance and maturity. This growth occurs in connection with certain symbols, including geometric symbols, and, if a person interacts with these symbols, drawing them, or meditating on them, this helps in the individuation process.

More precisely, each human being has four 'functions': thought, intuition, sensation, and feeling. One of these four, which varies from person to person, is unconscious, and needs to be integrated with the other three, to produce the unity which is the goal of individuation. This final state is symbolised by drawings which combine threes and fours into a single,

---

[11] Ibid., p. 20.
[12] Ibid., pp. 42-3.
[13] Ibid., p. 160.
[14] Ibid., p. 170.
[15] Ibid., p. 238.

balanced figure (e.g. by a drawing of a square divided into four triangles, each with a vertex at the square's centre).[16]

Several famous problems in Euclidean geometry involve a diagram which combines three things into a fourfold whole. For example, the problem of Apollonius is the problem of constructing a fourth circle tangent to three given circles. According to Jung, actually creating such diagrams, with ruler and compass, can help in the integration of one's personality. As one draws the straight lines and arcs joining the mathematical figures that symbolise the psychic components, the psychic components themselves are associated, in a way that leads to personal maturity.

Several famous Euclidean geometry problems do, indeed, involve three-four symbolism.

(a) Apollonius unites three given circles with a common tangent circle.

(b) The 'Problem of Napoleon' has us prove that the centres of three equilateral triangles, erected externally on the sides of a given triangle, are always the vertices of a fourth equilateral triangle.

(c) Collinearity theorems (such as Pascal's 'mystic hexagram' theorem) assert that three points are united in a single line.

There are also examples from other branches of mathematics. One of the Erdös problems is to show that any fraction of the form $4/n$ (with n an integer > 4) is a sum of three fractions with numerator 1. (For example, $4/7 = 1/3 + 1/6 + 1/14$.)

10.9 Whitehead's View

In *The Aims of Education and Other Essays*, A. N. Whitehead (1861–1947) advances a view of education which acknowledges both social and individual needs, and emphasises mathematics.

Whitehead rejects a sharp division between workers and managers. He feels that even the future workers should receive a 'liberal education'. This would include mathematics, not just as a 'training in concentration', but as part of an interdisciplinary understanding of the world.[17] If the workers are ignorant, they will be unhappy and they will not do their work as well. Thus

> alike for masters and for men [workers] a technical or techno-
> logical education, which is to have any chance of satisfying the

---

[16] C. G. Jung, *The Collected Works of C. G. Jung*, vol. 16, pp. 207, 317, and 321.

[17] A. N. Whitehead, *The Aims of Education* (London: Williams & Norgate, 1950), p. 35.

> practical needs of the nation, must be conceived in a liberal
> spirit as a real intellectual enlightenment in regard to principles
> applied and services rendered. In such an education geometry
> and poetry are as essential as turning laths.[18]

For Whitehead, the best way to train workers is not to habituate them
to mindless drill, but to give them a liberal education, including an understanding of mathematics. The future plumber should know why the
Theorem of Pythagoras is true, and he or she should know who Pythagoras was. Of course, Whitehead realises there are limitations. Many future
plumbers are neither willing nor able to learn much about, say, the relationship between ancient Greek society and the rise of modern science.
Hence Whitehead allows that, in teaching future workers, 'the theoretical
part should be clear-cut, rigid, short'.[19]

For the future managers, there is a greater need to learn mathematics.
The future manager needs to be able to be clear about the quantitative aspects of the world, and he or she must be able to reason and to generalise.[20]

Whitehead emphasises Euclidean geometry. Of course, there is more
than one way to teach it. Whitehead deplores the then current system of
having the students memorise a long chain of proofs for a final examination.
For Whitehead what counts is scientific understanding and the ability to
reason:

> Nobody can be a good reasoner unless by constant practise he
> has realised the importance of getting hold of the big ideas and
> of hanging on to them like grim death. For this sort of training
> geometry is, I think, better than algebra.[21]

Whitehead believes in the *Elements*.

> Euclid's fifth book is regarded by those qualified to judge as
> one of the triumphs of Greek mathematics. ...Nothing can be
> more characteristic of the hopelessly illiberal character of the
> traditional mathematical education than the fact that this book
> has always been omitted. It deals with ideas, and therefore was
> ostracised.[22]

---

[18] Ibid., p. 20.
[19] Ibid., p. 16.
[20] Ibid., pp. 11, 81-2, 127-8, and 135.
[21] Ibid., p. 128.
[22] Ibid., pp. 124-5.

10.10 A Plea for Euclid

With the possible exception of John of the Cross, every philosopher of education (whether in a broad or narrow sense of 'education') holds that mathematics is essential. Someone who knows about languages and history, but who has trouble with the basic properties of algebra, is simply not properly educated. Moreover, several thinkers stress the importance of studying Euclidean geometry. This is interesting because, about 1970, after having been studied for over two millenia, Euclidean geometry was abandoned, at least in the secondary schools and colleges of countries such as the United States. In this section we give some of the reasons for this apostasy, and urge repentance.

The fall of Euclidean geometry can be explained in terms of several factors.

First, it was often taught badly. As many teachers presented it, Euclidean geometry was just a long chain of proofs to be memorised for the exam.

Second, it was challenging. The problems of Euclidean geometry are real mathematics problems, not just rote applications of the 'usual procedure', and it is difficult to teach the creative, exploratory approaches needed to tackle real mathematics. Furthermore, there is the almost equally daunting matter of helping students express their constructions or proofs in good mathematical prose. It is so much easier for everyone if the students can get their marks by submitting one-number answers to routine formula questions!

A third reason for the demise of Euclidean geometry was the rise of 'new math'. Whereas professional mathematicians in the nineteenth century often did Euclidean geometry, professional mathematicians in the twentieth century devoted themselves to more abstract work. Set theory was a respectable subject, but the theory of the nine-point circle was considered jejune. For various reasons, it was felt that, even at the pre-university level, students would benefit more from doing twentieth century mathematics. They would have an inkling of what was going on in mathematical research, and they would be better prepared to do mathematical research themselves, if they wanted to. Hence Euclid was replaced by group theory, set theory, and logic.[23]

The 'new math' notwithstanding, Math scores on exams such as the Scholastic Aptitude Test kept dropping. Future plumbers had a better idea of what went on at the research level in algebra, but they still had trouble

---

[23] See J. A. Easley, 'Logic and Heuristic in Mathematics Curriculum Reform', in *Problems in the Philosophy of Mathematics*, ed. I. Lakatos (Amsterdam: North-Holland, 1967), pp. 208-30.

doing more concrete problems.

Those countries which eliminated Euclidean geometry from the secondary and college level curriculum lost something good.

First, they lost the training in reasoning which can be given by Euclidean geometry, properly taught, better than by any other branch of mathematics at the elementary level. Students who learned how to develop synthetic proofs, especially ones in which they had to discover their own 'construction', learned far more about reasoning and problem solving than 'follow-the-formula' students do.

Second, they lost touch with the past. Half the history of mathematics is Euclidean geometry, and the cultural heritage of humanity includes Euclid just as much as classical music. For centuries, Euclidean geometry influenced, not only science, but also literature and philosophy. No one can understand Aquinas, Newton, or Spinoza without having a grasp of the *Elements.* It is one of the keys necessary for opening the mind to world culture.

Third, in abandoning Euclid, they lost beauty. A Euclidean proof is often shorter, neater, more insightful, and more elegant than an analytic geometry proof of the same theorem. Where the students used to be exposed to the deft, limpid, vigorous demonstrations of Euclid, all they have now is the dull machinery of formulas. Even the future leaders must memorise a list of tricks for the exam. How can anyone create a new business, or propose a better health care scheme, unless they have imagination, and a sense of the apposite? How can anyone make intellectual discoveries for a new century unless they have learned to go beyond routine procedures?

In *Géométrie Classique et Mathématiques Modernes,* B. Sénéchal defends Euclidean geometry thus:

> Tool for understanding our world, source of various learning activities, simple and accessible model of the methods of mathematics, elementary geometry is an ideal subject for learning mathematics. It has, however, practically disappeared from the secondary level, as from the post-secondary. Despised by the majority of teachers because it is not a part of 'noble' theories, it deserves to be brought back.[24]

---

[24] B. Sénéchal, *Géométrie Classique et Mathématiques Modernes* (Paris: Hermann, 1979), p. 3. My translation.

10.11 Euclid and Ernest

For Paul Ernest, my defense of Euclidean geometry may put me in his 'old humanist' category, defending knowledge for the sake of knowledge, and not caring about how education might help improve society. To some extent this is true, but there is no reason why Euclidean geometry could not be studied in the conjectures-and-refutations style recommended by Ernest. The original Euclid is certainly not without faults, and there are many ways one could arrange his material. Euclidean geometry is by no means a unique deductive structure, sealed against intellectual debate.

10.12 Euclidean Geometry Today

Euclidean geometry is down but not out. People still do research in it, and, although they cannot publish this research in research journals, they can, and do, publish it in college level mathematics magazines. For example, Jordan Tabov published a new discovery in Euclidean geometry in the February 1995 issue of the *Mathematics Magazine.* The proof he gave for his discovery was not, alas, a Euclidean proof — he used complex numbers instead — but he none the less provided us with another beautiful truth in Euclidean geometry. Perhaps some college or secondary level teacher can encourage a student to find a wholly Euclidean proof for Tabov's theorem.

Tabov's theorem goes like this. Let PP' be a diameter of a circle, and let A, B, C, and D be any points on the circumference. If you drop perpendiculars from P to AB, BC and CA, the three feet of the perpendiculars are collinear in the 'Simpson' or 'Wallace' line for P and triangle ABC. The Wallace line for P and ABC is perpendicular to the Wallace line for P' and ABC, meeting it at the 'Griffiths point', on the nine-point circle of ABC.[25] Given a diameter PP', there is a Griffiths point for every triangle ABC whose vertices are on the circumference of the circle. What Tabov discovered is that the 4 Griffiths points for triangles ABC, ACD, ABD and BCD are collinear — on a line we should call the *Tabov line.*

There are many things a secondary or college level student might do in mathematics, but if he or she wishes to do something beautiful and original, if he or she wishes, not merely to memorise formulas, but to participate in the quest for knowledge, then I would recommend the problem of finding a purely elementary proof for Tabov's theorem. Just drawing the figure and laying a ruler across the four collinear Griffiths points would, I submit, be a far more valuable exercise than grinding out yet another routine problem in trigonometry or calculus.

---

[25] See R. A. Johnson, *Advanced Euclidean Geometry* (New York: Dover, 1962), p. 208.

Chapter 11

Mathematics and Religion

In this chapter we look at the interaction between theistic religion and mathematics. In the first nine sections we look at some of the implications of theism in general for the philosophy of mathematics, and in the last two sections we examine the relationship between mathematics and two specific theistic religions.

One critic dismissed a draft of this chapter on the grounds that it was a mere 'prayer'. Another thought it was 'dangerous' — that is, contrary to political correctness.

As for prayer, I am, indeed, exploring, in this chapter, the logical implications of the premisses of theism for various views within the philosophy of mathematics, but I am certainly not insisting that the reader accept those premisses, much less offer up a prayer. For example, when I talk about God's being able to create an infinite number of stars, I am not saying that there is a God and he has created an infinite number of stars (and the reader better thank Him for it). I am merely exploring the logical links between statements such as

There is a God

and statements such as

There cannot be an infinite quantity of anything.

As for political correctness, I realise that a certain type of academic 'liberalism' associates religion with intolerance, and becomes alarmed when religion is discussed in a way that is at all open to it. I also realise that, although this kind of academic liberalism pays lip service to 'academic freedom', it is not reluctant to take 'affirmative action' in the form of resisting the teaching or publication of whatever it classifies as 'incorrect'. My own view is that there are important connections between religion and mathematics and it is not inappropriate to examine them in a book such as this. Nor am I alone in holding this opinion. Volume 20 in Kluwer's *Episteme* series is a philosophy of mathematics monograph, written by S. Restivo, in which the author quite frequently discusses the relationship between religion and mathematics. In a section entitled 'Mathematics and God', Restivo gives examples which show that religion often has a positive effect on the progress of mathematics.[1]

In any case, it is my intention in this chapter, not to argue about whether I have a right to discuss religion in a philosophy of mathematics book, but simply to discuss it.

11.1 Infinity

As we noted in Chapters 1 and 2, the centuries have witnessed a long debate over the acceptability of the infinite. Even in the present, with Cantor's theory in the ascendant, there are thinkers like Edward Wette, who, writing in 1969, asserted that there are no integers greater than

$$10^{10^{10}}$$

From the point of view of the theist, finitism must be rejected. For, if the finitist is right, we have to picture God as not being able to create infinitely many stars, as not knowing infinitely many facts, and as not being able to love us for infinitely many years. In short, if the finitist is right, we have to picture God in a way that belittles him. Since this is wrong, it follows that there do exist those infinites on account of which we regularly praise God. God does have a knowledge of infinite patterns and sequences, and he is willing and able to share that knowledge with those creatures he has made in his image.[2]

---

[1] S. Restivo, *Mathematics in Society and History* (Dordrecht: Kluwer, 1992).

[2] Of course, there are non-theistic religions, like Buddhism, in which the good thing is not, say, having a friendship that lasts for infinitely many years, but having nothing at all. In this chapter, I am looking at religions in which one worships a unique, omniscient creator.

11.2 Platonism

One of the classical questions in the philosophy of mathematics is that of the ontological status of objects such as numbers, points or sets. The 'fictionalist' denies their existence outright, the 'materialist' claims they are really physical objects, and the 'intuitionist' accords them an existence only as finite concepts in human minds. The 'Platonist', however, holds that mathematical objects are both non-physical and independent of the human mind.

Now, since God is omniscient, he has a knowledge of mathematics. He knows, for example, whether Goldbach's Conjecture is true, that is, whether every even number greater than 2 is a sum of two primes. God knows this, moreover, because he is eternally aware of any counterexamples that might exist to this Conjecture. If there is a proof of Goldbach's Conjecture, then God certainly knows this proof, but, even if there is no proof — even if the Conjecture is 'undecidable' relative to the usual axioms of arithmetic — God none the less knows the truth of the matter, since he is fully aware of the contents of the set of sums of two primes, and he knows exactly which even numbers are missing from it.

In any case, as a consequence of God's knowledge of mathematics, it is false to say that mathematical objects exist only as concepts in the human mind. For they also exist in the mind of God. Indeed, even if God had chosen never to create a universe, he would still have known the truth about Goldbach's Conjecture, and hence numbers are not created, material objects. Nor do numbers depend for their existence on some human description of them. In short, theism implies that Platonism is true.

Against this, one might object that Platonism suggests that certain basic mathematical objects exist necessarily, and this does not fit in with the idea that God somehow 'creates' them, if only by thinking about them. To this objection, one can reply that a compatible necessity exists on God's side. It is necessary that God exist, and, if it is necessary that some mathematical object exist, this is only because it is necessary that God will its existence (in whatever sense of existence is being used here). God's will is perfectly free in relation to many different choices, but it is none the less constrained by his nature. He cannot, for example, choose to do evil. Nor can he choose to be ignorant of the correctness of Goldbach's Conjecture.[3]

Against the view that theism implies Platonism, one might also object that the Platonism it implies is pretty weak. If one holds that unicorns

---

[3] In all this, I am presupposing the orthodox view, championed by Aquinas, that (1) it is logically necessary that God exist and (2) it is logically necessary that there be at most one divine individual. If one adopts some heretical version of theism, such as 'moderate social trinitarianism', then one may end up in another place.

exist independently of human imagination, but only because God imagines them — without, however, creating them — then this hardly amounts to 'realism about unicorns'. Similarly, the fact that numbers exist in the mind of God does not mean they really exist. It merely means that, in discussing numbers as 'merely mental entities', one has to reckon also with the divine. To this objection, one can reply that, although there is more to a real unicorn than its being imagined, the number 3, for example, is no more or less than the content of someone's idea of it. Mathematical objects are the sort of thing that can exist in virtue a mind's awareness of them. If God wills the creation of a unicorn, something new appears in the universe, but if God wills the creation of the number 3, it simply continues to exist as an object of his thought.

11.3 Truth

As we noted in Chapter 5, Eli Maor has decided that, on account of the relative consistency of hyperbolic geometry, 'absolute truth had to be replaced by relative truth'.[4] Theists, however, reject relativism. For a theist, it is not merely relative that there is a god. Indeed, for a theist, there are quite a few 'absolute truths', including some absolute truths from mathematics.[5] For example, from a theistic viewpoint, it is absolutely true that

> God knows whether there is a sound proof of Goldbach's Conjecture

and it is absolutely true that

> God knows that $17 \times 45 = 765$.

The latter absolutely true statement implies yet others, such as

> 765 is a composite number

— and all these implications are necessarily true too. Indeed, if anyone maintains that, for him, 765 is not a composite number, then, far from his dissent giving evidence of the relativity of truth, it establishes merely that the dissenter is absolutely wrong.

---

[4] E. Maor, *To Infinity and Beyond*, p. 134.
[5] An absolute truth could be defined as a statement that is a precise and accurate description of an actual state of affairs. See Chapter 5.

11.4 Beauty

Yet another area in which theism has implications for the philosophy of mathematics is that of mathematical beauty. One implication of theism is that a proof should be judged beautiful, not merely in terms of the various marks of beauty — brevity, unification, excitement, and so on — but also in terms of the extent to which it expresses the divine. For the theist, a proof is beautiful insofar as it leads the soul to a closer acquaintance with the Logos, or Reason, of God.

Another implication of theism is that the most elegant line of research is also the most promising one. This is because God clothes with beauty only those things which also reveal his truth. There may be truth without beauty, but there is never beauty without truth.

11.5 History

According to some non-theists, history has demonstrated that religion is an enemy of Reason. Didn't the church try to silence Galileo?

The theist knows that this is not true. The theist knows that God created us with the ability to reason. God did this, moreover, because he wants us to have knowledge, not only about himself, but also about the beauties of creation. The history of mathematics is an intellectual journey thanks to which we can arrive at a deeper understanding of the Logos, and the patterns he has instantiated in the physical world.

The issue of Galileo is a distraction. However much it is described as a clash between Faith and Science, it was actually just a quarrel between proud wills. If a person is truly religious, he does not enter into power struggles. He submits to the truth, whether it is in religion or science. Because they refused to submit to the truth in science, Galileo's inquisitors were not acting in a religious fashion, they were not acting as true representatives of religion, and we cannot describe their actions as part of a war of religion against Science.

11.6 Education

If mathematics is to lead the student closer to God, the student has to be taught, not the rote learning of the computer, but the insight of the mystic. The student has to be taught to *think* about the subject. There must be marks for solving problems in a *non*-mechanical way. There must be marks for clever questions and fruitful suggestions. Every time a student uses a purely mechanical method to solve a problem, he or she should lose a mark — since, in that case, he or she is moving away from the Logos, and beginning to act like a machine.

Mathematics does include formulas and algorithms. Students should know why they work, and they should be able to program a computer to implement them. However, if students are to learn mathematics in a way that is sanctifying, rather than de-humanising, they should not spend much time carrying out rote procedures themselves. The 'assembly-line-method' may get you into the proudest university, but it will not lead you closer to the divine Truth.

In our present system of education, we divide everything up into specialised compartments. If you want Math, you go to the Math teacher, but if you want spiritual enlightenment, you go to the Chaplain. The result is a fragmented graduate. At church, he is the deepest of the faithful, but, at work, he is often just another mindless, rule-following bureaucrat. He still remembers the assembly-line-method for finding the maximum, but he still does not know why it works, or why there is a maximum, or why everyone is chasing after it.

11.7 Ethics

There are many ethical issues related to mathematics. Some of them are represented by questions such as the following.

(1) Should Professor Archimedes use his mathematical knowledge to help his government fight a war?

(2) Should he teach pure mathematics at a military college?

(3) Should the government give more money to mathematicians who are doing war-related research?

(4) Should the government take money from the hospitals and give it to number theorists?

(5) Should a clever person choose a career in medicine rather than in mathematics?

(6) Should one spend one's spare time on geometry, rather than helping the poor?

(7) Should one be a missionary rather than a mathematician?

For the theist, mathematics is important, but not as important as some other things. For the theist, then, there is no need to do or to teach mathematics at all costs, in all situations. The more important consideration is that of love. Will the mathematics hurt someone? Then do not do it!

11.8 Physics

Some people express surprise at the marvelous way in which mathematics applies to nature, giving us formulas, such as the Law of Gravity, which unlock its deep secrets. Why, they wonder, does something so abstract fit so well with the physical?

One answer is that mathematics deals with patterns. Now some pattern or other has to be instantiated in the physical world, and, whichever one it is, it belongs to the study of mathematics. If someone owned copies of every key in the world, it would not be surprising if, when he came to a door, he could eventually find the key that opened it.

A related answer is that the physical world is structured in terms of concepts (we can classify things), and concepts are structured in terms of mathematics. For example, as Frege points out in *The Foundations of Arithmetic,* each concept has a finite or infinite number belonging to it, namely, the number of things falling under that concept. The fact that mathematics applies to nature is thus no more or less surprising than the fact that concepts apply to nature.

The theist has a third, possibly more satisfying answer. For the theist, the reason mathematics applies to nature is that God built it into nature. God 'ordered all things by measure, number and weight' (Ws 11:20). Thus it is no more surprising to find that a certain kind of mathematics applies to creation than it is to discover that a certain 'blueprint' describes a building. The gap between the 'invisible' and the 'visible' is bridged by the divine intellect.

11.9 Number Worship

Theists often maintain that, although a study of numbers can help lead one to the ultimate reality, the ultimate reality itself is not a number, or a pattern, but a thinking, willing person. The final destiny of humanity depends, not on the rate of the expansion of the universe, not on the equations in the theory of entropy, but on the freely made decisions of God.

This version of theism presupposes the view that there is an important difference between numbers, patterns, or laws, on the one hand, and persons, personal decisions, and divine providence, on the other. Thus if one worships a number, or system of numbers, one is committing idolatry.

In her fascinating book *What Number is God?*, S. Voss entertains the possibility that God's being a person does not preclude his also being a mathematical entity.[6] It may seem like a category mistake to say, for example, 'God is his oneness', but when we consider that any talk about God is necessarily unequal to its referent, and when we consider that some great orthodox thinkers, such as Aquinas, have said similar things, we may wish to take a second look at Voss's position.

Some university employees take their numbers very literally, and this, together with their belief that academic reasoning is more important than the insights of the saints, has led some of them into heresy. We have, for example, the 'moderate social trinitarians' who, finding it too difficult to distinguish the literal from the analogical in the traditional doctrine of the trinity, have decided to simplify matters by becoming polytheists. However, suppose someone — not a professor of philosophy, but a real philosopher — claims that he or she is both a Christian and a Pythagorean. The point Voss is making is this. Yes, sure, you can defend some definitions of 'number' and 'person' from which you can deduce that the Christian Pythagorean is illogical and therefore bad and to be ostracised and not given a letter of recommendation. However, it might be more fruitful to ask, 'why can a number not also think and make decisions?' Why should numbers generally be 'mere abstractions' rather than something more commensurate with the important role they play in determining the limits of the possible?

Pythagoreans hold that numbers have a reality of their own. The number 10, for example, is as real as this book. Moreover, just as books have properties over and above their text — for example, they can be heavy or green — so numbers have properties over and above their mere mathematical content — for example, certain numbers symbolise perfection or friendship. In a suitably Pythagorean context, there need be nothing either illogical or heretical in declaring that the number one is omniscient — or that 'God is his oneness'.

In 1996 I attended a talk by Roger Penrose, after which someone asked if he believed in a personal God.[7] He replied that he believed in a Platonic 'Goodness'. This was not the reply the questioner was hoping for, but, given the inadequacy of human language for the task of expressing the nature of the divine, I am not sure we should rush to the conclusion that Penrose is 'unsaved'. Aquinas, at any rate, is rejoicing that Penrose, already fully

---

[6] S. Voss, *What Number is God?* (Albany: State University of New York Press, 1995).
[7] The talk was given on May 25, at 10 Merton Street in Oxford.

understanding the first part of the *Summa Contra Gentiles,* now needs only to study Book IV.

11.10 Islam

History of mathematics books sometimes have a chapter on 'Islamic mathematics', but they never have a chapter on 'Jewish mathematics' or 'Christian mathematics'.[8] Is there something, then, about Islam that relates it to mathematics in a special way?

Before answering this question, we should note that the term 'Islamic mathematics' is itself a misnomer. For one thing, there is nothing particularly Islamic about the theorems, so that, for example, a Christian would be in danger of heresy if he accepted them. For another, the mathematicians usually cited in this connection are not all of them Muslims: Thabit Ibn-Qurra, for example, was a Sabian. Finally, the term 'Islamic mathematics' is usually applied to mathematics done in the middle ages, and thus has the incorrect connotation that Muslims no longer do mathematics in the modern era.

This having been said, there is still the question of whether Islam and mathematics are related in some special way.

Ali Al-Daffa' thinks they are. In his book *The Muslim Contribution to Mathematics,* he points out the following links between Islam and mathematics.

(1) Muslims need geometry to calculate the direction to Mecca.

(2) The Koran encourages Muslims to 'study the sky and the earth to find proofs of their faith'.[9]

(3) A 'great religious impulse' caused Muslims to create a civilisation, one that included many mathematical achievements (such as the development of trigonometry).

Al-Daffa' is right about these things, and they do indicate that there is a relationship between Islam and mathematics. They do not, however, bear on the possibility that other religions have a similar relationship to mathematics. The Christian, for example, needs arithmetic to calculate the date

---

[8] This is true, for example, of Victor Katz's book *A History of Mathematics.*

[9] Quoted by Ali A. Al-Daffa' in *The Muslim Contribution to Mathematics,* (London: Croom Helm, 1977), p. 9.

of Easter. The Bible tells him to 'prove all things' (1 Th 5:21). A 'great religious impulse' caused Christians to create a civilisation, one that included many mathematical achievements (such as the development of calculus and complex analysis). Indeed, many of the great western mathematicians were fervent believers in Jesus.

What is special about Islam in connection with mathematics is that it is an extraordinarily abstract religion. God is totally invisible and totally transcendent. In some religions you can have a talk with God, you can bring him round to your point of view, you can relate to him as to a brother. In Islam, you simply submit, and adore God as the one and only totally other. Islam has an austerity and other-worldliness which is mitigated in most other religions. And this austerity and other-worldliness fits very well with the austerity and other-worldliness of some kinds of mathematics. Unable to put pictures or statues in their mosques, Muslims decorate them with geometric patterns. Unable to locate God in a concrete human person, Muslims seek him in the abstract patterns behind his creation. Unable to find God in anything merely finite, Muslims seek him in the infinite.

Needless to say, this is an oversimplification. There are versions of Islam which offer very concrete ways of relating to God — just as there are extremely other-worldly, Platonic forms of Christianity. On the whole, however, Islam is the more purely transcendent religion, and it therefore provides a more natural philosophical home for someone who studies the transcendent in pure mathematics. If mathematics continues its trend to ever greater abstraction, and if Islam continues its current revival, we shall see some important contributions to mathematics from future Islamic scholars.

11.11 Christianity

Christianity is sometimes portrayed as an anti-mathematics religion, and there were, indeed, some so-called Christians who were against mathematics. In *The History of the Church*, Eusebius quotes an 'orthodox' writer as follows:

> They put aside the sacred word of God, and devote themselves
> to geometry — earth-measurement — because they are from
> the earth and speak from the earth, and do not know the One
> who comes from above. Some of them give all their energies to
> the study of Euclidean geometry ....[10]

---

[10] Eusebius, *The History of the Church*, trans. G. A. Williamson (London: Penguin, 1965), p. 177.

This is not, however, a typical Christian view — if it is a Christian view at all. The typical Christian view is pro-mathematics. Origen, for example, believed that mathematical activity is helpful in coming closer to God. Eusebius praises Origen in these words:

> First he taught them geometry, arithmetic, and the other prepara-
> tory subjects; then he led them on to the systems of the philoso-
> phers ....[11]

Indeed, in every age, there have been famous Christians who endorsed the study of mathematics. Bishop Anatolius (d. 282) wrote a book on number theory. Augustine (d. 430) promoted the pro-mathematical views of Plato. Pope Sylvester II (d. 1003) wrote a mathematics book. Saint Edmund of Abingdon (d. 1240) taught mathematics at Oxford. Bishop Oresme (d. 1382) produced original results on infinite series. And there are examples of pious Christian mathematicians in more recent times: Pascal, Leibniz, Cauchy, Cantor, and so on.

Granted, then, that Christianity favours mathematics, does it favour any particular kind of mathematics? Christianity is a more concrete religion than, say, Islam, and it is interesting to note that much 'Christian mathematics' is applied mathematics. For example, they were mostly Christian mathematicians who developed the calculus and worked out its first applications in the natural sciences.

Of course, Christianity also has its other-worldly side, and it is not surprising that the founder of transfinite set theory was a Christian.

Christianity is a religion based on the two-natured Christ. Corresponding to Christ's divine nature, Christianity has a 'vertical' aspect, an aspect which can express itself in contemplative prayer — or in transfinite set theory. Corresponding to Christ's human nature, Christianity has a 'horizontal' aspect, an aspect which can express itself in care for the sick — or in applied mathematics.

---

[11] Ibid., p. pp 194-5.

Chapter 12

Mathematics and Reference

Communication involves referring to things. When we communicate, one of our goals is to get the persons with whom we are communicating to realise what it is we are referring to. We expect them to be looking for a referent. They expect us to be seeking a way of helping them pick out just whatever it is we have in mind. Usually this is done by means of language, but there is always also pointing or pantomime. I may not know your language, but, if I dress like a tourist, walk into your restaurant, point to an item of food on display, and take out my wallet, I shall certainly be communicating with you.

As players of charades know, there is something different about referring to things other than medium sized physical objects. 'Dead dog' is going to be easier to act out than 'prime numbers'. Indeed, even with the help of language, it can sometimes be difficult to get someone to realise that what you have in mind is prime numbers. They may, for example, have trouble understanding the difference between prime numbers and odd numbers, and you may have to do a lot more than pointing to sort things out.

In this chapter we work on the problem of referring to mathematical objects (that is, mathematical structures and their components). How is it done (if it is done)? And what makes it different from other kinds of referring? These are the sorts of questions we deal with.

12.1 Reference and Language

One of the advantages of having a well-developed language is that it allows you to refer to things quickly and easily. A well-developed language is a sophisticated communication tool, and its effectiveness is a sign that its vocabulary and its grammar in some way 'match up' with the content or the purpose of the communication. This match up may not be perfect, but it has to be good enough to meet basic needs. A group of shepherds using a single word to refer to baby sheep and adult sheep may not run into difficulties, but a group of shepherds with only one word for both dog and wolf might soon see the advantages of introducing new vocabulary. Against a general scepticism about language and reference, I should like to assert that, insofar as a language is effective, it does allow us to refer to things as they really are, independent of our language. I grant that qualification is needed — the 'reality' of things is here conditioned by the needs and goals of the language speakers — but I do not grant that the needs and goals of the language speakers are unrelated to brute facts in a brute world in which we must survive.

12.2 Crying Wolf

Suppose there is just one wolf in the district. We know of no others. Suppose I want to warn you that the wolf is coming. So I call 'wolf!'. You are supposed to realise that I am not merely exercising my lungs (why would I keep looking at you in that case?) but that I am trying to draw your attention to something. To what? To the greyish shape moving across the meadow? Not only. To the greyish shape that resembles that other greyish shape which, as we both well remember, ate one of your lambs last week? Yes, you have caught on! I am not merely referring to the north surface of a time-slice of the wolf, but to a real wolf, with four sides (as it were), and an invisible but dangerous appetite. There might be a word for the north side of a wolf spotted at noon, and there might be a word for something that was either the wolf or the dog, but an effective language, for shepherds at least, ought to have a word for that particular object the wolf.

The wolf is really, in a sense, an abstract object. It is an abstraction of its various concrete manifestations at various particular times. Indeed, the fact that there is a single word for this thing is the conclusion of a whole range of experiences of the wolf, correlating its appearance, its speed, and its appetite.

We might imagine some early scientists talking:

> do you realise that those speedy grey things all make the sheep disappear?

— But it's really just one thing that keeps coming back, and it's different from the other thing that only eats already dead sheep.

— Let's call the live sheep-eater the 'wolf' and the other one the 'hyena'.

The point is this. We refer to a reality that is independent of us, but in a way that serves our interests. We do not, therefore, refer to a raw or unprocessed version of reality. We abstract away the details of little interest and we organise together what seems important. But we do refer to reality.

12.3 Abstraction, Classes, and Equivalence Classes

If there are some objects we can refer to, and they can be collected together (in thought) as having some common feature, then we can refer to the abstract object which is that collection, getting the person with whom we are communicating to pick out that collection by means of guiding his or her attention to typical or average members of the collection.

For example, if there are now several wolves in our district, we shall soon be referring, not to the wolf, but to the more abstract 'any wolf'.

Furthermore, given a collection that we can refer to, we can partition its elements into equivalence classes on the basis of certain features shared within each equivalence class. For example, the collection of visible things can be partitioned up in terms of the seven colours of the rainbow. Yellow is just the equivalence class of all yellow things, and if we want to refer to yellow, we can point to some yellow objects that have little in common except their colour.

A collection of equivalence classes can itself be partitioned up into equivalence classes, giving us abstract objects at a higher level of abstraction, giving us, say, polygonal-shaped objects as opposed to square-shaped objects. As another example, start with the collection of grey patches. We can partition it up into dangerous patches, friendly patches, and neutral patches. We can then partition the resulting three element collection into a collection of only two elements: important grey patches and neutral grey patches.

This process of abstraction is commonplace in mathematics. The numbers modulo 6, for example, are abstracted from the natural numbers by ignoring any differences between two natural numbers if they leave the same remainder when divided by 6.

As for the wolf, he is an abstraction of certain sense data concerning the white shapes disappearing into the grey shapes. By ignoring the time, the weather, and so on, we arrive at the beast himself.

Since the wolf appears at a very low level of abstraction, we call him 'concrete', but it is none the less true that the abstraction process has already begun.[1]

The reference problem of mathematics is not one of referring to abstract objects per se — even the colour yellow and the wolf are abstract objects in some sense — but of referring to objects at a very high level of abstraction. We have to get our interlocutor's mind to follow us up to the right level of abstraction, and this will typically involve referring to equivalence classes of equivalence classes of etc.

12.4 Reference and Hale

In *Abstract Objects,* Bob Hale raises an objection to the idea of taking (some) abstract objects to be equivalence classes.[2] What if, by chance, the only kind of ink we know about is purple ink, and things written or drawn in this ink are the only purple things in our experience? The various instances of purple will make up the equivalence class P which is the abstract object purple. The various instances of ink will make up the equivalence class I which is the abstract object ink. But $P = I$, and hence the abstract object purple is the abstract object ink, which seems wrong.

To this one can reply, first, that it would create no problems in mathematics. In mathematics we are quite happy to identify sets with the same members, regardless of how each is defined, and we even identify sets that are merely isomorphic. Natural numbers are just sums of four natural number squares? Good!

More to the point, perhaps, is the fact that Hale's sort of counterexample is unstable. In the situation envisaged, we could easily enlarge our experience by imagining the result of splashing the purple ink on a white shirt to produce a new kind of object for the class of purple things. If the identity of equivalence classes P and I is merely a chance identity, not fixed in any way by reality, then it will not survive long enough to be a barrier to communication.

12.5 Reference and Precision

In order to refer to something it is not necessary to do so in an unambiguous way. The genuineness of the reference depends on the success of the

---

[1] It follows from this that we can have a posteriori knowledge of abstract objects, provided the level of abstraction is low enough — as it is in the case of a wolf, or a word in Chinese, or a geometric diagram.

[2] B. Hale, *Abstract Objects* (Oxford: Basil Blackwell, 1987), p. 185.

communication, not some independent criterion of precision. If I tell you about

> a man taking a walk in Central Park at two in the morning

you will understand me to be referring to a state of affairs in which the man is not part of a crowd. After all, you realise that I realise that, given our common background assumptions, you will take it this way. So, although I do not explicitly specify that fact that the man is not in a crowd, you will not object to my saying

> If a man takes a walk in Central Park at two in the morning, he will be robbed.

You will not object to this by pointing out

> But if he is escorted by fifty policemen, he will not be robbed.

We both realise that if I wanted to refer to *that* situation, I ought to have said so. Since we know we are communicating, there is no need for complete precision in my words. Indeed, if you know me well enough, I may only have to grunt, in a certain context, to make you realise you should pass the marmalade.

According to the definite description theory of reference, what is referred to is what the words actually describe. If you say

> the third prime number

then you are referring to 5. This theory is wrong because it does not take into account the history or context of your words — this including common knowledge of common beliefs. You and I might just have learned about prime numbers from the famous Professor Smith who mistakenly listed the first three as

> 3, 5, 7

After our class, we talk about the 'third prime number' and we are both thinking about 7, not 5. Mathematicians can be careless about their words, and still succeed in talking to each other. The logic of the problem, the shared insights, and so on, fix the reference, at least to the extent that genuine communication is possible. Any lack of precision does not undermine the mathematical work in progress.

Related to this is the issue of first and second order logic, a distinction most mathematicians do not bother with, but which is of great concern to their critics. Those who go for first order descriptions of mathematical theories have to deal with the fact that first order theories admit 'nonstandard models' as possible referents. Those who go for second order descriptions of mathematical theories have to deal with the fact that second order theories admit arbitrary infinite subclasses, not subject to any defining principle. In either case, there is less than perfect clarity, less than total precision. None the less mathematicians do refer to their theories. I will not be misunderstood if, in discussion with a mathematician, I refer to 'the natural numbers'.

Perhaps it is like colours. By 'yellow', I mean strong, solid yellow. By 'yellow', you mean any kind of yellow, and you are willing to call yellow even a surface with a few small nonstandard red dots on it. None the less, I shall know what you mean if you say,

> Look at that yellow sportscar!

— even if it is only pale yellow. Part of effective communication is recognising, and compensating for, a certain fuzziness in word use.[3]

12.6 Reference and Inscrutability

Sceptics exaggerate the fact that we sometimes talk at cross purposes. They warn us about the possibility of an 'inscrutability of reference', thanks to which we may always be talking at cross purposes — and yet never be able to know it. You say

> Wilt is tall

and I understand you to mean just that, but what you actually have in mind is, say,

> Wilt's shadow is long, even at three o'clock

— and we never realise we are talking of different things.[4] For the sceptic, this sort of situation is, not the exception, but the norm. Our language (and knowledge) are simply not 'sharp' enough to cut up reality in a unique, unambiguous way. We have certain 'models' of reality, but we cannot ever be sure which one is the most accurate.

One reply to the sceptic is that, if one wants to engage in philosophical dialogue in a way that is not purely destructive, one has to presuppose that, in communication, there is a 'meeting of minds'. This 'meeting' is

---

[3] Note that we are discussing vagueness in language, not vagueness in reality. To say that our words fail to pick out a single kind of set does not imply that sets are fuzzy objects. It merely implies that there are different things that might be sets.

[4] D. Davidson, *Inquiries into Truth and Interpretation*, (Oxford: Clarendon Press, 1984), p. 230.

not based merely on our imperfect language, or on our empirical knowledge, but on a capacity we human beings have for intuiting, to some extent, the contents of each other's inner thoughts. This may seem mysterious, but it is also commonplace, and the evidence for this capacity is simply our frequent success in guessing what another person 'has in mind', in spite of the inadequacies of our knowledge, or of our articulation of it.[5]

12.7 Reference and Ontology

Following Husserl, I claim it does not matter if the referent does not exist.[6] For, as puzzling as it may seem, we do regularly refer to nonexistent things.
Consider the sentences

Meinong's round square? It does not exist.

The pronoun 'it' clearly refers to the round square. Or consider the question

Are you referring to Mickey Mouse?

This reasonable question might have a correct, affirmative answer.

People can and do refer to whatever they have in mind, regardless of its ontological status (which may even be entirely unknown). And other people understand what is meant.

In mathematics, one quite often refers to impossible objects. Consider, says the mathematician, a rational number equal to the square root of two. Then, following a little reasoning, the mathematician proclaims: 'but now we have a contradiction, and so the rational number we have been talking about cannot possibly exist'.

Here one might ask, if one can refer to non-existent objects, what is the difference between referring to Sherlock Holmes and referring to Einstein? There is no difference in the referring. In both cases one is simply drawing the attention of one's interlocutor to what one has in mind. Indeed, one could refer to Sherlock Holmes under the mistaken impression that he existed, exactly as one could refer to Einstein under the mistaken impression he never existed. The difference is not in the referring but in the fact: something in reality corresponded to Einstein but nothing in reality corresponded to Sherlock Holmes.

Of course, it is usually the case that one refers to something as existing or as nonexisting. That is, usually one has an opinion about the existence

---

[5] For the theist, there is the additional consideration that God would not create people just to have them live in the darkness of misunderstanding. He made us, in his image, with minds sufficiently powerful to perceive something of the true nature of reality, and also something of the true meaning of other people's sentences.

[6] D. Bell, *Husserl* (London: Routledge, 1990), p. 129.

of one's referent, and one lets one's interlocutor know if one believes that the referent does not exist. The absence of any indications that the communicator doubts the existence of the referent is taken to be an indication that the communicator does not doubt the existence of the referent. This is a convention of communication which anyone wanting to communicate will make use of. For effective communication, I must let you know if I mean to be 'pretending'.

12.8 Reference and Numbers

From looking at various packs of wolves of various sizes, we arrive at the notion of 'pair of wolves', and, similarly, we arrive at the notion of 'pair of sheep', and so on. Then, with more abstraction, we arrive at the notion of 'pair of like objects', and finally at the general notion of 'pair' and the number 2.

In order to refer to the number 2, it is not necessary to have a well-developed language. When the Chinese write

$$-, =, \equiv$$

they are using expressions which undercut language and are therefore understood by those who cannot read Chinese. To understand that what the Chinese are writing are numbers, it is enough to realise that the marks are deliberately simple, and deliberately similar, the idea being that what is referred to are the bare cardinalities, rather than any other qualities of the marks.

This pre-axiomatic kind of reference to small whole numbers is in play when we talk about the 'standard' model of whatever fancy axiom system we are using for the positive integers. It is partly what makes our reference to the standard numbers possible. Some philosophers might object to this, wanting reference to mathematical objects always to be totally divorced from anything 'concrete' or 'empirical'. However, as indicated above, even the wolf presupposes some abstraction, and the move to the more abstract need not involve any radical discontinuity.

12.9 Reference and Large Numbers

What about numbers like 12^{60}? We obviously do not abstract them out of some experience with large packs of wolves. None the less we do refer to such numbers, and this on the basis of the number system. Given the small positive integers, we organise them according to the so-called laws of arithmetic to form a system, including exponentiation, which allows us to refer to 12^{60}.

The laws of arithmetic are generalised out of our experience with small numbers, an experience which is partly spatial. The distributive law, for example, is understood in terms of a diagram with two adjacent rectangles.[7]

Of course, if reference proceeds by arithmetical laws, then a full explanation of reference must incorporate an explanation of our understanding and use of those laws. How is it we look for them, understand them, use them — and refer to them?

Suppose I define an operation for positive integers this way:

$$n^1 = n$$

$$n^{m+1} = n^m \times n$$

Somehow you will see what I mean, and in such a way that you will be able to calculate 12^{60}. Somehow you will grasp the fact that what I am communicating is like a general set of instructions, and you will know how to apply them in specific cases. To do this, you have to have some idea of what a general set of instructions is, and you have to realise I am trying to refer to such a thing. Perhaps one could sketch a cognitive theory about how students learn exponentiation from doing lots of calculations and how they learn to interpret arbitrary recursively given rules by starting with those that define operations they already know via lots of calculations. Moreover, it is not impossible that such a theory would have to refer to linguistic abilities that are innate and which somehow guide the abstraction process, leading some students to catch on, while others consistently fail to understand. Perhaps such a theory, to be successful, would have to refer to some capacity human beings have for an 'intuitive understanding'.

12.10 Reference and Benacerraf

Benacerraf raised the following question.[8] Do the numerals

$$0, 1, 2, \ldots$$

refer to the abstract objects

$$\phi, \{\phi\}, \{\phi, \{\phi\}\}, \ldots$$

or to the abstract objects

$$\phi, \{\phi\}, \{\{\phi\}\}, \ldots$$

---

[7] One has a 3 by 3 rectangle followed by a 3 by 5 rectangle to illustrate the equation $3(3 + 5) = 3 \times 3 + 3 \times 5$.

[8] P. Benacerraf, 'What Numbers Could Not Be', *Philosophical Review* 74 (1965).

or to some other sequence? And if we cannot answer this question, does
that mean they do not refer to any abstract objects at all?

On the analysis of abstraction suggested above, the numeral n refers to
the class C_n of all sets that can be put into one-to-one correspondence with
the first n natural numbers.[9] For example, 2 is the class of pairs. This
is an idea going back to Russell.[10] And if Russell is right, the answer to
Benacerraf's question is that the numerals refer to the sequence C_0, C_1,
C_2, and so on.

Another way of answering Benacerraf's question is simply to say 'all
of the above'. That is, we could take it that the numeral 2 refers to the
class K_2 of all ordered pairs of the form $(f(SS0, M), M)$ where M is any
sequence of abstract objects and $f(SS0, M)$ is the second nonzero member
of it.[11]

Note that we do not have to answer Benacerraf's question by saying
that we cannot refer to numbers, and we do not have to answer Benacerraf's
question by saying that numbers are fuzzy objects.

12.11 Reference and Azzouni

In *Metaphysical Myths, Mathematical Practice,* J. Azzouni considers an
argument that goes something like this.[12]

> If one could refer to numbers, one could, not just mix up the
> terminology we use for them, but even mix *them* up.
> But one cannot.
> So one cannot refer to numbers.

As an example of mixing things up, Azzouni points out that we can mix
up identical twins. I think that is John the mathematician but actually it
is James the botanist. In support of his second premiss — the 'infallibility'
premiss — Azzouni gives various examples, claiming that, although one
could mix up twins, one would not mix up, say, 5 and 7 (though, thanks
to an incompetent professor, one might think the numeral 7 refers to the
number 5).

Against Azzouni, I believe one can mix up 5 and 7. It is just that it is not
easy. It is like confusing a cannon-ball and a bowl of soup — the problem is

---

[9] In order to avoid 'impredicativity' (see below), one can exclude those sets whose
members are natural numbers, or whose members are 'built up' from natural numbers.

[10] See page 18 in Russell's *Introduction to Mathematical Philosophy.*

[11] For yet another answer to Benacerraf's question, see the last chapter of Bob Hale's
Abstract Objects.

[12] J. Azzouni, *Metaphysical Myths, Mathematical Practice* (Cambridge: Cambridge
University Press, 1994), pp. 46-47.

not lack of reference but the fact that a cannon-ball is so evidently different from a bowl of soup. Consider, however, the case of Johnny who is just learning to count. He says

$$1, 2, 3, 4, 5, 6, 5, 6, 7, 8, \ldots$$

— No, no, Johnny! After 6 comes 7. You must keep counting up and not ever start to count down.

— But last night Susy told me I was lucky since I got 5 cookies and she only got 6.

I claim that Johnny has mixed up 5 and 7.

Moreover, as Azzouni himself admits, mathematics is usually not infallible. If someone can make a mistake of kind k in the empirical sciences, then, probably, he or she can make a mistake of kind k in mathematics.

12.12 Reference and Impredicativity

Suppose I refer to $\sqrt{2}$ as the least upper bound of the sequence whose n-th term is $\sqrt{2}$ to n decimal places. My reference to $\sqrt{2}$ is via a set of upper bounds, including numbers like 3. However, this set, like any set, is determined by its members, including $\sqrt{2}$. So my reference to this set involves an implicit reference to $\sqrt{2}$. We are going in a circle. My reference is 'impredicative' in the sense that I have referred to an entity in terms of a totality of which it is a part.

If one understands a reference to a mathematical object — at least an intitial reference to it — as somehow creating that object, then one might well conclude that the circle is vicious. How can you erect the beam on the floor which is to be supported by the beam? However, if one understands the reference as merely picking out an already given object — possibly imaginary but none the less given — then there is no problem. If you succeed in picking out the item of interest for your interlocutor then the reference is a success.

Indeed, one form of impredicative reference is usually taken to be, not just inoffensive, but canonical. I am talking about the impredicative definitions typically given for plants and animals. The ninth edition of the *Concise Oxford Dictionary* defines a rhododendron as

a shrub ... esp. an evergreen one, with large clusters of bell-shaped flowers.

The reference to rhododendron goes via a reference to shrub, and it goes in spite of the fact that one could not have a complete understanding of 'shrub' without knowing one's rhododendrons. According to some reference books, then, there is nothing untoward about impredicative reference.

In mathematics, some thinkers (such as Russell) have been suspicious of impredicative definitions because they allow things like the Russell Paradox, and they seem to commit one to Platonism. However, the paradoxes can be barred in other ways than by outlawing impredicativity, and, as we noted above, one can refer to objects that are in some sense given, without committing oneself to their ontological status. There seems to be no compelling philosophical reason on the basis of which one might insist that mathematicians give up their rather common practice of making impredicative references.

12.13 Reference and Infinity

Suppose I write

$$|, ||, |||, ||||, \cdots$$

I am referring to some collection of numbers, but which one? A finite but elastic one with no particular end? A completed actual infinity? Often, we can 'fix a reference' for a collection whose members can already be referred to by pointing to some typical members. By the 'upper form clique', I mean people like that Tilly Vavasour. In the case of our sequence, however, what numbers should I point to? Suppose I point to 999 and 10^6. That may not help. For what if someone has trouble believing I might be referring to an actually infinite set, and thinks that I mean something like 'numbers that can be written down using less than a page of paper'? Perhaps I can publish a counting schedule according to which the number n gets counted at time $1 - 1/2^n$, and then claim that I am referring to all the numbers that would have been counted by time 1. But suppose my interlocutor does not understand just what set I have n ranging over. Then what? Is my interlocutor stupid, or I am just kidding myself that I am referring to some actually infinite collection? Perhaps I am just referring to some huge collection of numbers which will serve me for all the practical mathematical purposes I could possibly imagine, but which is still finite.

Let us try this. I am referring to as many natural numbers as it needs 9's to make

$$0.999999\ldots = 1$$

But my interlocutor is not sure that any number of 9's will do. I can get 1 with arbitrary precision, but is it really possible that I get exactly 1? If, at every stage I shall be short, why should I get the full 1 at some nonstage which is the infinite?

Let us try again. We draw the attention of the person with whom we are communicating to finiteness. We talk about sets S such that any one-to-one function $f : S \rightarrow S$ is onto. Then we say, think of sets that do not have this quality, that is, sets S such that there is a one-to-one correspondence $f : S \rightarrow S$ which is not onto S. However, what if our interlocutor complains that he cannot imagine what this function f is? Yes, he can imagine 0 going to 1, 1 going to 2, and 'so on', but what does the 'so on' mean? Is there an $\omega + 4$ going to an $\omega + 5$? Does mere negation create reference, with no given 'universal set' with respect to which the negation is taken? What, after all, does 'not Mickey Mouse' refer to?

Lavine tackles the infinite with the help of the expressions 'extrapolation' and 'simplificatory rounding out'.[13] However, although he presents an elaborate theory of the finite — this is supposed to be the jumping off point for the extrapolation — he says little about the extrapolation itself. We are supposed to 'extrapolate' by forming

> a single domain of quantification V that is so large that it can never require enlargement in any context.[14]

However, it is not explained how we are to conceive of this domain V. All the domains previously considered by Lavine were domains requiring enlargement. Nor is it certain that what we would be conceiving would be actually infinite as opposed to being, say, unimaginably large, but still finite. Lavine's reader is left wondering if 'extrapolation' is not just some mysterious power which might, yes, be pointing to the actually infinite, but which might equally well be assimilating the infinite to the indefinitely large — confusing the two notions.[15] Lavine himself seems rather unsure of the outcome of his 'extrapolation':

> any length we can imagine could be prolonged in some larger context. We can therefore have no grasp on just how long unimaginably long is.[16]

If what Dummett says is right, we should give up trying to communicate the idea of the infinite. He asserts:

> no compelling refutation can ever be offered of a radical finitism whose proponents purport not to understand what it means

---

[13] S. Lavine, *Understanding the Infinite*, p. 251.
[14] Lavine, p. 316
[15] Lavine, p. 289-90.
[16] Lavine, p. 317.

to say that there are infinitely many things of a certain kind, whether natural numbers or stars.[17]

Yet even a radical finitist can study set theory and make intelligent conversation about various kinds of infinite set. What is going on?

Let us start with mathematics as it seems to be practised. Mathematicians do, in fact, use a notion of the actual infinite, they do refer to it, and they do communicate views about it. There may be a certain ambiguity or vagueness in their language, but mathematicians none the less succeed in helping each other prove theorems about the infinite, theorems which they all accept.

Since, as we showed above, it does not seem easy to explain reference to the infinite merely in terms of some 'extrapolation' from finite data, we might try some other form of explanation. One such was offered by Descartes, and this is that the infinite is an innate idea.

What is an innate idea? For Plato it would be a memory of a Form. For Jung it would be a psychic archetype. For a Jew or Christian, it would be a component of the 'image of God'. For a materialist, it would be a gene — useful somehow for the purposes of survival. No matter how it is viewed, an innate idea is something which guides us in conceptualising the world, and which guides us in how we ought to expect others to be conceptualising the world when they communicate with us. The radical finitist Dummett has in mind either lacks this innate idea (being handicapped) or else lacks access to this innate idea (being repressed). Yes, perhaps for ideological reasons, or perhaps out of what Cantor called the 'fear of the infinite', we can repress a clear and distinct idea of it.

In Chapter 3 we argued in favour of the infinite. Our arguments were important in the sense that it is this openness, or lack of it, that is a key factor in one's philosophy of mathematics. Those who are open to the infinite will be able to accept a mathematician's saying something like

> Either Goldbach was right about the infinite set of evens, or he was not

whereas those who are not open to the infinite will have to interpret this as confusion. On the view defended by Dummett there is no innate idea to guide our minds to the infinite, and mathematicians, for all their precise definitions, are in the dark about the infinite. It is as if a group of cooks liked to discuss their recipes in terms of an incoherent theory of alchemy.

> You add the yeast and knead the bread and a mercury of pure tinsel raises its wings in the heat of the October snore.

---

[17]M. Dummett, 'The Source of the Concept of Truth' in *Meaning and Method,* ed. G. Boolos (Cambridge: Cambridge University Press, 1990), pp. 14-15.

Ah! But for raisin bread, it is the November snore.

On the view expressed by Dummett, mathematicians are, in effect, spending a lot of time producing babble.[18] However, if mathematicians are making sense, if they are really communicating, then it is possible to refer to the infinite, and what Dummett says is wrong.[19]

It is very easy to pick something out and be sceptical about it. It is harder to be sceptical about everything, or, if one is not going to be sceptical about everything, to defend one's choice of *this* rather than *that* as the object of sceptical attack. Dummett promulgates a kind of selective scepticism of the infinite, but another person, with different presuppositions, could equally easily attack the notion of the finite. The selective sceptic has only to announce that he 'does not understand' whatever it happens to be — daring anyone to come and *make* him understand — but, meanwhile, actual mathematicians are doing hard and successful work analysing every possible sort of concept, including the infinite. Indeed, mathematicians quite regularly, and quite naturally, take a 'bird's eye view' of their activities, 'looking down' on infinite collections of mathematical operations and the objects that go with them. If Dummett wants to single this out as a target of doubt, and to denigrate it as mere nonsense, then he is only demonstrating that he does not appreciate the noble Spirit of the mathematical adventure.

[18] M. Dummett, *The Elements of Intuitionism* (Oxford: Oxford University Press, 1977), p. 360.

[19] For a refutation of Dummett, see G. Hunter's article 'Dummett's Arguments About the Natural Numbers', *Proceedings of the Aristotelian Society,* 80 (1979/80), pp. 115-126.

Appendix A

Suggestions for Further Thought

For Chapter 1

1. Did Zeno prove that there is an infinite?

2. What do you make of Aristotle's arguments against the actually infinite?

3. Give the details in Aristotle's second argument against the infinite. For example, write down the formula which tells us exactly where the point P is at a time t between 3 and 4 o'clock, and another formula telling us its instantaneous velocity at that time. Work out P's average velocity, and draw a displacement versus time graph, and also a velocity versus time graph for P's motion. What is P's acceleration at 3:13 PM? Explain how the point P can always have finite instantaneous velocity, and yet have infinite average velocity.

4. Is the following definition of the potentially infinite good or bad? Why?

   > A collection of objects is potentially infinite if its size is perfectly elastic, able to stretch to any finite dimension, no matter how large.

How would *you* give a precise definition of the potentially infinite?

5. Is a theist committed to the view that there are actual infinities in mathematics?

6. What did different medieval philosophers in the Islamic tradition have to say about the infinite in mathematics?

7. A Zeno type argument can be given to show that the real numbers cannot be listed in a sequence

$$r_1, \ r_2, \ r_3, \ r_4, \ \ldots$$

For suppose such a sequence contains all the reals. Then the set of real numbers is a subset of the infinite union of real line intervals

$$\left[r_1 - \frac{1}{2}, \ r_1 + \frac{1}{2}\right] \cup \left[r_2 - \frac{1}{4}, \ r_2 + \frac{1}{4}\right] \cup \left[r_3 - \frac{1}{8}, \ r_3 + \frac{1}{8}\right] \cup \cdots$$

These intervals 'cover' the set of reals. However, whereas the total length of the real line is infinite, the total length of the above union is not greater than

$$1 + \frac{1}{2} + \frac{1}{4} + \frac{1}{8} + \cdots \leq 2$$

Contradiction. Comment on this argument.

For Chapter 2

1. What was Pascal's view of the infinite. Why did he hold this view?

2. Summarise and comment on the anti-infinity arguments of the empiricists. Would modern set theory have made them change their minds?

3. G. W. F. Hegel (1770–1831) attempted to answer Berkley's objection to Leibniz's use of dx by saying that dx is neither zero nor nonzero. Like Aristotle, Hegel felt that two truth values are not always enough. Comment. (See Hegel's *Science of Logic*, trans. A. V. Miller (London: George Allen & Unwin, 1969), pp. 104-5, 137, 254-5, 269-70, 438-9.)

4. Kronecker said, 'God Himself made the whole numbers — everything else is the work of humans'. Comment.

5. Cantor said:

> The fear of infinity is a form of myopia that destroys the possibility of seeing the actual infinite, even though it in its highest form has created and sustains us, and in its secondary transfinite forms occurs all around us and even inhabits our minds.[1]

Comment.

6. On page 115 in the April 1995 issue of *Scientific American*, A. W. Moore suggests that the undecidability of the Continuum Hypothesis means that 'we should rethink how well Cantor's work tames the actual infinite'. Would Moore say that undecidable statements in ordinary arithmetic mean that we should rethink how well Peano's work tames ordinary arithmetic? Comment.

7. Suppose someone argues like this: on Cantor's view the power set of any set S has a greater cardinality than S. Thus the power set of the set X of all sets has a greater cardinality than X does, and yet it is a subset of X. This contradiction means Cantor was wrong. Rebut this argument.

8. Was Wittgenstein a finitist? What reasons did he give for the position he actually held?

9. Suppose that at time $1 - 1/2^n$ we put 10 balls in an urn and then, at time $1 - 2/(2^n 3)$, we withdraw one ball at random (for $n = 1, 2, 3, \ldots$). Show that the probability that the urn is empty at time 1 is 1.

   Hint: let p_n be the probability that a particular ball, going into the urn at time $1 - 1/2^n$, is removed at some stage; then

   $$p_n = \frac{1}{10n - (n - 1)} + \frac{10n - n}{10n - (n - 1)} p_{n+1}$$

---

[1] R. Rucker, *Infinity and Mind* (Boston: Birkhäuser, 1982), p. 43. Rucker is quoting Cantor.

For Chapter 3

1. Pick one of the Chapter 3 arguments, and rebut it.

2. Comment on the following pro-infinity argument, with special attention to its use of the law of the excluded middle:

    Suppose there are only finitely many natural numbers.
    Then there is a largest one, n.
    But $n + 1$ is also a natural number, and larger than n.
    Contradiction.
    Thus, by reductio ad absurdum, it follows that there are infinitely many natural numbers.

3. Comment on the following pro-infinity argument, explaining how it might be rebutted:

    If a line segment is not made up of infinitely many points, but only a finite number, then each of those points will have some finite length $\epsilon > 0$.
    So every line will have a length of the form ϵn, where n is a natural number.
    Thus, without loss of generality, we can suppose that the legs of an isosceles right triangle each measure ϵa, where a is a natural number.
    By the Theorem of Pythagoras, the hypotenuse measures $\epsilon a \times \sqrt{2}$.
    Also, given our assumption, this hypotenuse measures ϵb, for some natural number b.
    Hence $\epsilon a \times \sqrt{2} = \epsilon b$.
    Thus $\sqrt{2} = b/a$ — a rational number.
    Contradiction.

4. Present the four arguments of Zeno as pro-infinity arguments.

5. Create your own pro-infinity argument, distinct from those above.

6. Suppose someone wants to reserve the word 'infinite' for God. Define 'transfinite' to mean 'not finite'. Hence 'transfinite' describes not only God but also the set of counting numbers. Can the arguments

in this section be adapted to show that human beings are transfinite? Explain.

7. According to Christian teaching, God became finite in Jesus — so it is possible for a person to be both infinite and finite (in different respects). According to the teachings of some other religions, God's power is infinite in such a way that he lacks the power to become finite. Comment.

For Chapter 4

1. How would you defend your preferred ontological school?

2. How can the if-thenist reply to the objection that the if clauses are, for him, not only false, but necessarily false, and hence all the if-then statements are necessarily, but trivially, true?

3. In *Unreality*, C. Crittenden claims that it is logically impossible for a fictional object to exist. Thus if squares are fictional objects, they do not differ in this respect from round squares — both are impossible. Comment.

4. Aristotle talks about the impossible, the contingent and the logically necessary — three 'modalities of being'. (An object X is *contingent* just in case it is possible that X exist and also possible that X not exist. For example, Diana is contingent, but round squares are not.) Could a realist hold that mathematical objects are contingent? Could a nihilist hold that mathematical objects are contingent? If the nihilist holds that mathematical objects are not contingently nonexistent, but impossible, is there any valid way in which he or she might separate statements about these impossible objects into true and false?

5. What should a structuralist say if someone asks about the ontological status of the structures? (On page 34 of his 'Mathematical Knowledge and Pattern Cognition' (*Canadian Journal of Philosophy* 5 (1975)) Michael Resnik claims that structuralism is a form of Platonism. Is

this right?)

6. Would an intuitionist be well advised to defend that position by saying that branches of mathematics not open to intuitionists are not important?

7. How does Maddy defend concretism?

8. What should a formalist say if someone asks about the ontological status of the 'collection of numerals for 2'?

9. Would a Platonist be well advised to defend that position by saying that we can imagine an endless grid of squares (as on graph paper), and, in so doing, we are 'perceiving' the immaterial Euclidean plane?

10. Is Platonism undermined by the fact that we do not know which set the number 2 is? (For example, we could just as well identify it with the set $\{\{\phi\}\}$ as with the set $\{\phi, \{\phi\}\}$.) Should the Platonist say that 2 is the class of all sets that might stand in for 2? Or should she say that 2 is simply the 2 created by God — which need not be a set at all?

11. Summarise M. Balaguer's isolationist Platonism, described in his article 'A Platonist Epistemology' (*Synthese* 103 (1995), 303-25). Is Balaguer's view that mathematical objects are vague essential to his position? Given that all possible mathematical objects can be modeled within ZFC (with sufficiently large cardinals), does Balaguer really need 'all possible structures' — as opposed to just ZFC (with sufficiently large cardinals)?

12. In *Foundations of Arithmetic,* Gottlob Frege (1848–1925) eschews idealism as 'psychological', and writes

    even the mathematician cannot create things at will, any more than the geographer can; he too can only discover what is there and give it a name.

    Report on Frege's thoughts on the ontology of numbers.

13. Report on Adam Drozdek and Tom Keagy's 'A Case for Realism' in volume 77 (1994) of the *Monist.*

14. What are the fundamental particles of mathematics? Are they symbols, sets, natural numbers, points, functions, fuzzy sets, truths of logic, types, all of the above — what?

15. How have different philosophers and educators used the terms 'construction' and 'constructivist'?

For Chapter 5

1. What is the Poincaré model of hyperbolic geometry, and how does it relate to the issue of truth?

2. What is meant by the expressions 'model' and 'non-standard interpretation'?

3. What is Poincaré's 'conventionalism'? Is it a form of pragmatism? Is it true? (See H. Poincaré, *Science and Hypothesis* (New York: Dover, 1952), pages xxv and 41.)

4. Argue that, indeed, hyperbolic geometry is too weird to be true. Do this in the context of various standard theories of 'truth'.

5. In 1963, Frederic Fitch proved that if verificationism is true, then we are omniscient. Comment on his argument, which goes as follows. Let p be any true proposition. To obtain a contradiction, suppose we do not know that p. Then the following proposition q is true:

   (1) p and (2) it is not the case that we know that p.

   According to verificationism, for any proposition r,

   r only if it is possible that we know that r.

   Hence if verificationism is true,

it is possible that we know that q.

Hence it is possible that, at the same time, (1) we know that p, and (2) we know that it is not the case that we know that p. Since (2) implies that it is not the case that we know that p, it follows that it is possible that, at the same time, (1) we know that p, and (2') it is not the case that we know that p. Contradiction.[2]

6. Does Eli Maor commit an 'all-or-nothing fallacy' when, from the premiss that the Parallel Postulate is not an absolute truth, he concludes that there is no absolute truth at all? (In your answer, define this fallacy and give other examples of it.)

7. Can we compare a revolution in mathematics with a rejection in love? If the shocked mathematician concludes that no theorem is ever to be counted on ever again, is this like the jilted lover concluding that no member of the opposite sex is ever to be trusted ever again? In the case of love, at least, this is seen as an over-reaction. Comment.

For Chapter 6

1. In 1993, Andrew Wiles announced that he had a proof of Fermat's Last Theorem. Great embarassment followed when a gap was found in the proof, but, happily, Wiles was able to repair the damage, and present a valid proof the year later. Dig up the details of this event, and comment on it.

2. When I was teaching at Duffer College, I asked my students to prove that the radius joining the centre of a circle to the point where it meets one of its tangents is perpendicular to that tangent. One student tried this argument:

> Let P be the point of tangency, and let d be a line perpendicular to the radius. Let Q be any other point on line d. Then, if A is the circle centre, APQ is a right triangle. Since the hypotenuse of a right triangle is the longest side, $AQ > AP$ and thus Q is not on the circumference. So P is the only point on d which meets the circle. Hence d is a tangent to the circle at P. Since d is perpendicular to the

---

[2] *Topoi* 13 (2) (1994) contains several articles on Fitch's argument.

radius at P, the tangent is perpendicular to the radius at the point of contact. QED.

Say why this argument is not a good argument. What are the gaps? What went wrong?

3. I asked my students to prove that if two circles are tangent then their two centres and the point of tangency are collinear. Some merely drew a picture, and noted that the line through the circle centres passed through the point of tangency. This sort of proof is not rigorous because it says nothing about tangent circles in general: it only addresses the special case of the theorem where the radii are exactly the radii of the circles on the drawing. It is not rigorous anyway, because it relies on the accuracy of the drawing, rather than on reasoning. If the pencil were sharper, perhaps the line would *not* pass through the point of tangency.

Another error in the 'proofs' was to assume that the circles must touch externally. No one even attempted to deal with the case in which one centre is in the interior of the other circle. This was another example of 'jumping to conclusions'.

One attempted 'proof' went this way:

Let A and B be the circle centres and P the point of contact. Each of the radii AP and BP is perpendicular to the common tangent at P (see previous assignment problem answered above). Since two right angles add to a straight angle, AP and BP lie in a straight line.

Why is this not a rigorous proof?

4. Give perfectly rigorous proofs of the above two theorems.

5. Suppose AB and CD are two parallel chords in the same circle. Then $\angle ACD = \angle BDC$. There is a very short proof of this, but, in trying to find it, students often jump to various conclusions. For example, they will assume, without warrant, that the given chords are equal, or that the diagonals of the quadrilateral whose four vertices are A, B, C, D meet in the centre of the circle, or that B, D are on the same side of AC (giving only one of the two diagrams). Students will give long proofs involving two or more triangle congruences. Or they

will confuse the fact that the interior angles between two parallels are supplementary with the fact that the opposite angles in a cyclic quadrilateral are supplementary. And they will confuse alternate angles with angles standing on the same arc. I once marked about thirty such proofs! Write an essay explaining how one might prepare one's students for such a question, so that they do not submit unrigorous proofs.

6. The June 1995 issue of *Scientific American* contains an article on 'complexity' in which we learn that at least 31 definitions of that term have been proposed. Is complexity really so complicated? Do all of these definitions apply in the case of mathematics?

7. On page 51 of his book *Undergraduate Algebraic Geometry*, Miles Reid writes:

> The Zariski topology may cause trouble to some students; since it is only being used as a language, and has almost no content, the difficulty is likely to be psychological rather than technical.

What is the Zariski topology? Is it an example of worthless abstraction? Should Reid have omitted it?

8. On page 61 of his *Philosophical Introduction to Set Theory,* Stephen Pollard writes:

> We perform abstraction when we restrict our language in such a way that a relatively weak relation is able to impersonate identity. The idea is that we pick out — or abstract — those forms of expression which do *not* allow us to see through the impersonation. The remaining forms of expression are "left behind" as residue.

What exactly does Pollard mean? Is he right?

9. In Book VI of Wordworth's *Prelude* he writes about geometry:

> Mighty is the charm
> Of those abstractions to a mind beset
> With images and haunted by herself,
> And specially delightful unto me

> Was that clear synthesis built up aloft
> So gracefully; even then when it appeared
> Not more than a mere plaything, or a toy
> To sense embodied: not the thing it is
> In verity, an independent world,
> Created out of pure intelligence.

What was Wordsworth's view on the ontological status of geometric objects? What was his view on elegance in mathematics? Was there a connection between the two?

10. In 'Notes on the Value of Science', Lars Bergström complains that 'a large percentage of the scientific work that is published is bad or boring'.[3] He also quotes Quine's statement that new journals 'were needed by authors or articles too poor to be accepted by existing journals'. In mathematics, at any rate, one could cut down on the number of useless publications by insisting that the proposed contributions be meaningful as well as valid. Comment.

11. Does excessive complexity imply meaninglessness? (See page 43 in *Fuzzy Logic* by Daniel McNeill and Paul Freiberger.)

12. In his address at Paris 1900, David Hilbert claimed that rigour leads to simplicity and fruitfulness:

> The very effort for rigor forces us to discover simpler methods of proof. It also frequently leads the way to methods which are more capable of development.

Comment.

For Chapter 7

1. Pick a famous theorem, such as the Theorem of Pythagoras, and rate two or more proofs of it, and then compare the ratings.

---

[3] See p. 506 in *Logic, Methodology and Philosophy of Science IX*, ed. D. Prawitz et al. (Amsterdam: Elsevier Science, 1994).

2. Is an elegant proof elegant in all possible universes? Is it necessarily elegant?

3. If it is a necessary truth that A is better than B, is it, not just wrong, but absolutely wrong, to replace A by B, with no overriding reason for doing so?

For Chapter 8

1. Pick one of the Chapter 8 questions, and give your own response to it.

2. What about the view that an historian should simply write about what interests him or her, without any regard to philosophical values?

3. Write a short review of a history of mathematics textbook, commenting on the author's prejudices and presuppositions.

4. Find out what we really do know about Hypatia. Make a list of the things people have made up about her.

5. Why do writers include Galileo in history of mathematics textbooks? Did he do any original mathematics? Or is he just there so the author can make some philosophical point?

6. Should we use fuzzy logic to evaluate the 'authorship' of mathematical discoveries?

For Chapter 9

1. Is there a counter-example to the 'way of knowing' definition of infallible knowledge?

2. Is it self-evidently true that $13 \times 13 = 169$, or is this statement too complicated to be self-evident?

3. If a triangle is divided by a straight line into two parts, then the sum of the areas of the two parts equals the area of the triangle. Is this a self-evident statement? Does it follow from the mere definition of area? Which definition of area? In general, how can you tell if a self-evident statement is self-evident?

4. Comment on Robert Audi's definition of a statement's being self-evident: understanding it is sufficient for being justified in believing it and believing it on the basis of understanding it is sufficient for knowing it.

5. Consider someone who wants to have absolutely certain knowledge of mathematical truths. Why do they want this? Do they suffer from a neurotic insecurity? Do they suffer from hubris? Do they want to challenge God, taking for themselves something that rightly belongs only to him?

6. Visit a research seminar in Mathematics and report on how they come to know things in mathematics.

7. A statement is analytic just in case its truth or falsity can be deduced from the meanings of its words. Otherwise it is synthetic. For example, the following statement is analytically true:

> If the union of 5 sets has 6 elements then at least one of those 5 sets contains at least 2 elements.

Is every analytically true statement a logically necessary statement? Is every logically necessary statement an analytically true statement? What about the Power Set Axiom, or the Axiom of Infinity?

8. Defend some examples of a priori knowledge of synthetic truths. Are there any examples of a posteriori knowledge of analytic truths? Why?

9. In *Science and Hypothesis*, Henri Poincaré writes that mathematical induction,

inaccessible to analytical proof and to experiment, is the exact type of the *a priori* synthetic intuition. On the other hand, we cannot see in it a convention as in the case of the postulates of geometry.

Comment.

10. Are there any self-evident statements that are not analytic? According to Audi's definition of self-evidence, are all analytic statements self-evident?

11. What would Plato say about a posteriori knowledge in mathematics? What would he say about the geometrical examples of a posteriori knowledge (given above)?

12. Compare the computer-humanist distinction given above with the distinction between experience-intensive and logic-intensive knowledge given in Michael Detlefsen's article 'Brouwerian Intuitionism' in *Proof and Knowledge in Mathematics* (London: Routledge, 1992).

13. What is the relation between 'humanist reasoning' and fuzzy logic?

14. Suppose God has created the material universe in such a way that it cannot be fully understood except in terms of a mathematics which endorses the Continuum Hypothesis. Suppose that, eventually, physicists do understand it this way. Could we then be said to have a posteriori knowledge that the Continuum Hypothesis is true? How would Plato react to this idea?

For Chapter 10

1. Pick one of the philosophies of education discussed in Chapter 10, and comment on it, in relation to the role of mathematics.

2. What did Rousseau say about mathematics in education?

3. How much rigour should Math students be required to master? Why?

4. Find a Euclidean proof of Tabov's theorem — so you will know what it is like to do original work in mathematics, so you will be able to inspire others to do the same.

5. No one should teach mathematics, at any level, unless he or she is capable of *doing* mathematics. Comment.

6. In an attempt to have the computer emulate human thought, researchers are trying to create programs that do synthetic Euclidean geometry. The programs are not yet entirely successful, but impressive work has been done by Chou, Gao, and Zhang. Report on their book *Machine Proofs in Geometry*.

7. Write a report on Philip J. Davis's article 'The Rise, Fall, and Possible Transfiguration of Triangle Geometry' (in the March 1995 issue of the *American Mathematical Monthly*).

For Chapter 11

1. The Axiom of Choice is true if there is a God who can make the choices. Comment on this statement.

2. Comment on the following argument.

> Necessarily, God exists.
> Necessarily, if God exists then all the natural numbers exist
> (at least as objects of his awareness).
> Hence, necessarily, all the natural numbers exist.
> Thus, in all possible worlds, there is an infinite collection.
> Hence, theism implies that the Axiom of Infinity is a truth
> of logic.

3. The Bible says that 'the plan of the city is perfectly square' (Rv 21:16). Since there are no squares in hyperbolic geometry, could this be taken to imply that hyperbolic geometry is not the geometry of

heaven?

4. Get up a list a great mathematicians, together with their religions, and their mathematical specialities. Are there any interesting correlations?

5. Archimedes killed Roman soldiers. A Roman soldier killed Archimedes. This was perfectly just and logical. As Jesus says, 'all who take the sword will die by the sword' (Mt 26:52). Comment.

6. List some famous Hindu mathematicians, and say something about the relationship between Hinduism and openness to the infinite.

7. What is the relation, if any, between Buddhism and mathematics?

8. The Japanese sometimes carve geometric diagrams into the stone of their temples. Give an example of this, and explain why it is done. (See H. Fukagawa and D. Pedoe, *Japanese Temple Geometry Problems*, Winnipeg: The Charles Babbage Research Centre, 1989.)

9. In order to be fair, a writer of a textbook on the history of mathematics should not have a chapter on 'Islamic Mathematics' unless he or she also has a chapter on 'Christian Mathematics' and a chapter on 'Jewish Mathematics'. Comment.

For Chapter 12

1. When we refer to the noncyclic group of order 4, are we referring to a structure which has its own structure, or what?

2. What about our references to groups in the abstract? Is an abstract group a group? If it is, how many elements does it have? If it is not, why do we talk about it in group theory?

3. In category theory one identifies isomorphic objects. For example, in the category of sets, any singleton set is a 'terminal object'; they are all isomorphic. We might say that the singleton is the terminal

object. What is this terminal object? Are we referring to the class of all singletons? To some abstract singleton? To the number 1 — as suggested by category theory notation?

Appendix B

Bibliography

Anglin W. S., *Free Will and the Christian Faith* (Oxford: Clarendon Press, 1990).

— *Mathematics: A Concise History and Philosophy* (New York: Springer-Verlag, 1994).

— *The Queen of Mathematics* (Dordrecht: Kluwer, 1995).

— and J. Lambek, *The Heritage of Thales* (New York: Springer, 1995).

Aquinas, T., *Summa Theologiae* (London: Blackfriars, 1964-1981).

Archimedes, *The Works of Archimedes*, ed. T. L. Heath (New York: Dover, 1897).

Aristotle, *The Collected Works of Aristotle*, ed. J. Barnes (Princeton: Princeton University Press, 1984).

Aschenbrenner, K., *The Concepts of Value* (Dordrecht: Reidel, 1971).

— and A. Isenberg (eds.), *Aesthetic Theories* (Englewood Cliffs: Prentice-Hall, 1965).

Augustine, *The City of God*, trans. Henry Bettenson, ed. David Knowles (New York: Penguin Books, 1972).

Azzouni, J., *Metaphysical Myths, Mathematical Practice* (Cambridge University Press, 1994).

Baker, A., and H. Davenport, 'The Equations $3x^2-2 = y^2$ and $8x^2-7 = z^2$', *Quarterly Journal of Mathematics*, ser. 2, 20 (1969), 129-137.

Ball, W. W. R., *A Short Account of the History of Mathematics* (New York: Dover, 1960).

Bar-Hillel, Y. (ed.), *Logic, Methodology and Philosophy of Science* (Amsterdam: North-Holland, 1965).

Barnes, J., *Early Greek Philosophy* (London: Penguin, 1987).

— *Pre-Socratic Philosophers* (London: Routledge and Paul, 1979).

Bell, D., *Husserl* (London: Routledge, 1990).

Bell, E. T., *The Development of Mathematics,* 2nd edn. (New York: McGraw-Hill, 1945).

— *Men of Mathematics* (New York: Simon & Schuster, 1937).

Benacerraf, P., 'What Numbers Could Not Be', *Philosophical Review,* 74 (1965), 47-73.

— and H. Putnam (eds.), *Philosophy of Mathematics,* 2nd edn. (New York: Cambridge University Press, 1983).

Benardete, J. A., *Infinity* (Oxford: Clarendon Press, 1964).

Bishop, E., and D. Bridges, *Constructive Analysis* (Berlin: Springer-Verlag, 1985).

Boethius, *Boethian Number Theory,* trans. M. Masi (Amsterdam: Rodopi, 1983).

Bolzano, B., *Paradoxes of the Infinite,* trans. Fr. Prihonsky (London: Routledge and Kegan Paul, 1950).

Boolos, G. (ed.), *Meaning and Method* (Cambridge: Cambridge University Press, 1990).

Boyer, C. B., and U. C. Merzbach, *A History of Mathematics,* 2nd edn. (New York: John Wiley, 1989).

Brindza, B., 'On Some Generalisations of the Diophantine Equation $1^k + 2^k + \cdots + x^k = y^z$', *Acta Arithmetica,* 44 (1984), 99-107.

Brown, J. R., *The Laboratory of the Mind* (London: Routledge, 1991).
Burton, D., *The History of Mathematics,* 2nd edn. (Dubuque: William C. Brown, 1991).

Chihara, C. S., *Constructibility and Mathematical Existence* (Oxford: Clarendon Press, 1990).

Chowla, S., rev. of 'Contributions to the Theory of Diophantine Equations', by A. Baker, *Mathematical Reviews,* 37 (1969), 737.

Comte, A., *Cours de Philosophie Positive* (Paris: Anthropos, 1968).

Conche, M., *Anaximandre: Fragments and Témoignages,* (Paris: Presses Universitaires de France, 1991).

Copleston, F., *A History of Western Philosophy,* (New York: Doubleday, 1985).

Côté, A., 'Aristote admet-il un infini en acte et en puissance?', *Revue Philosophique de Louvain,* 88 (1990), 487-503.

Coxeter, H. S. M., and S. L. Greitzer, *Geometry Revisited* (Washington: The MAA, 1967).

Daffa, A., *The Muslim Contribution to Mathematics* (London: Croom Helm, 1977).

Dauben, J. W., 'Cantorian Set Theory and Limitations of Size', *British Journal for Philosophy of Science,* 39 (1988), 541-550.

— *Georg Cantor* (Cambridge, Mass.: Harvard University Press, 1979).

Davidson, D., *Inquiries into Truth and Interpretation* (Oxford: Clarendon Press, 1984).

Davis, M., 'Hilbert's Tenth Problem is Unsolvable', *American Mathematical Monthly,* 80 (1973), 233-269.

Davis, P. J., and R. Hersh, *The Mathematical Experience* (Boston: Birkhäuser, 1981).

Descartes, R., *The Geometry,* trans. D. E. Smith (New York: Dover, 1954).

— *The Philosophical Works of Descartes,* trans. E. S. Haldane and G. R. T. Ross (New York: Dover, 1953).

Dummett, M., *The Elements of Intuitionism* (Oxford: Oxford University Press, 1977).

— *The Logical Basis of Metaphysics* (Cambridge, Mass.: Harvard University Press, 1991).

— 'Truth', in *Truth,* ed. G. Pitcher (Englewood Cliffs: Prentice-Hall, 1964).

— *Truth and Other Enigmas* (Cambridge: Harvard University Press, 1978).

Dunham, W., *Journey through Genius* (New York: John Wiley, 1990).

Elkies, N. D., 'On $A^4 + B^4 + C^4 = D^4$', *Mathematics of Computation,* 51 (1988), 825-835.

Ellison, W. J., and F. Ellison, J. Pesek, C. E. Stahl, and D. S. Stall, 'The Diophantine Equation $y^2 + k = x^3$', *Journal of Number Theory,* 4 (1972), 107-117.

Ernest, P., *The Philosophy of Mathematics Education* (London: The Falmer Press, 1991).

Euclid, *The Elements,* trans. T. L. Heath (New York: Dover, 1956).

Eusebius, *The History of the Church,* trans. G. A. Williamson (London: Penguin Books, 1965).

Everett, E., *Orations and Speeches* (Boston, 1870).

Eves, H., *An Introduction to the History of Mathematics,* 5th edn. (Philadelphia: Saunders, 1983).

Ewing, J. H., *Numbers,* trans. H. L. S. Orde (New York: Springer-Verlag, 1990).

Ferreirós, J., 'Traditional Logic and the Early History of Sets', *Archive for History of Exact Sciences,* 50 (1996), 5-71.

Field, H., *Science Without Numbers* (Princeton: Princeton University Press, 1980).

Finkelstein, R., and H. London, 'On D. J. Lewis's Equation $x^3 + 117y^3 = 5$', *Canadian Mathematical Bulletin,* 14 (1971), 111.

Franklin, J., 'Achievements and Fallacies in Hume's Account of Infinite Divisibility', *Hume Studies,* 20 (1994), 85-101.

Gauss, C. F., *Disquisitiones Arithmeticae,* trans. A. A. Clarke (New Haven: Yale University Press, 1966).

George, A., 'How Not to Refute Realism', *Journal of Philosophy,* 90 (1993), 53-72.

Gettier, E., 'Is Justified True Belief Knowledge?', *Analysis,* 23 (1963), 121-123.

Gillies, D. (ed.), *Revolutions in Mathematics* (Oxford: Clarendon Press, 1992).

Grabiner, J. V., 'Is Mathematical Truth Time-Dependent?', *American Mathematical Monthly,* 81 (1974), 354-365.

Greenberg, M. J., *Euclidean and Non-Euclidean Geometries* (San Francisco: W. H. Freeman, 1974).

Grünbaum, A., *Modern Science and Zeno's Paradoxes* (Middletown: Wesleyan University Press, 1967).

Hale, B., *Abstract Objects* (Oxford: Basil Blackwell, 1987).

— 'Is Platonism Epistemologically Bankrupt?', *Philosophical Review,* 103 (1994), 299-325.

Hallett, M., *Cantorian Set Theory and Limitation of Size* (Oxford: Clarendon Press, 1984).

Hardy, G. H., *A Mathematician's Apology* (Cambridge: Cambridge University Press, 1967).

— 'Mathematical Proof', *Mind,* 38 (1929), 1-25.

Heath, T., *A History of Greek Mathematics* (New York: Dover, 1981).

Hilbert, D., *Foundations of Geometry* (Chicago: Open Court, 1971).

Hintikka, J. (ed.), *The Philosophy of Mathematics* (London: Oxford University Press, 1969).

Hobbes, T., *The English Works of Thomas Hobbes* (London: John Bohn, 1839).

Hollingdale, S., *Makers of Mathematics* (London: Penguin Books, 1989).

Hume, D., *A Treatise of Human Nature,* ed. L. A. Selby-Bigge (Oxford: Clarendon Press, 1896).

Hunter, G., 'Dummett's Arguments About the Natural Numbers', *Proceedings of the Aristotelian Society,* 80 (1979/80), 115-126.

Huntley, H. E., *The Divine Proportion* (New York: Dover, 1970).

Ireland, K., and M. Rosen, *A Classical Introduction to Modern Number Theory* (New York: Springer-Verlag, 1982).

John of the Cross, *The Collected Works of St. John of the Cross,* trans. K. Kavanaugh and O. Rodriguez (Washington: ICS Publications, 1979).

Johnson, L. E., *Focusing on Truth* (London: Routledge, 1992).

Johnson, R. A., *Advanced Euclidean Geometry* (New York: Dover, 1962).

Jung, C. G., *The Collected Works of C. G. Jung,* trans. R. F. C. Hull (New York: Pantheon Books, 1960).

Kanagasabapathy, P., and T. Ponnudurai, 'The Simulataneous Diophantine Equations $y^2 - 3x^2 = -2$ and $z^2 - 8x^2 = -7$', *Quarterly Journal of Mathematics,* ser. 2, 26 (1975), 275-278.

Kanigel, R., *The Man Who Knew Infinity* (New York: Charles Scribner's Sons, 1991).

Kant, I., *Critique of Judgement,* trans. J. H. Bernard (New York: Hafner, 1951).

— *Critique of Pure Reason,* trans. Norman Kemp Smith (London: Macmillan, 1929).

Katz, V. J., *A History of Mathematics* (New York: HarperCollins, 1993).

Kershner, R. B., 'On Paving the Plane', *American Mathematical Monthly,* 75 (1968), 839-844.

— and L. R. Wilcox, *The Anatomy of Mathematics* (New York: Ronald Press, 1950).

Kirk, G. S., and J. E. Raven, *The Presocratic Philosophers* (Cambridge: Cambridge University Press, 1957).

Kirkland, L., *Theories of Truth* (London: MIT Press, 1992).

Kitcher, P., *The Nature of Mathematical Knowledge* (New York: Oxford University Press, 1983).

Kline, M., *Mathematical Thought from Ancient to Modern Times* (New York: Oxford University Press, 1972).

— *The Loss of Certainty* (Oxford: Oxford Unversity Press, 1980).

Koblitz, A. H., *A Convergence of Lives* (Boston: Birkhäuser, 1983).

Kripke, S., *Naming and Necessity* (Cambridge: Harvard University Press, 1980).

Lakatos, I. (ed.), *Problems in the Philosophy of Mathematics* (Amsterdam: North-Holland, 1967).

Lambek, J., and P. J. Scott, *Introduction to Higher Order Categorical Logic* (Cambridge: CUP, 1986).

Lang, S., *Fundamentals of Diophantine Geometry*, 2nd edn. (New York: Springer-Verlag, 1983).

— 'Mordell's Review', *Notices of the AMS*, 42 (1995), 339-350.

Lavine, S., *Understanding the Infinite* (Cambridge: Harvard University Press, 1994).

Lebesgue, H., *Measure and the Integral*, ed. K. May (San Francisco: Holden Day, 1966).

Leff, G., *Gregory of Rimini* (Manchester: Manchester University Press, 1961).

Lehman, H., *Introduction to the Philosophy of Mathematics* (Totowa: Rowman and Littlefield, 1979).

Lehrer, K., *Theory of Knowledge* (Boulder: Westview Press, 1990).

Leibniz, G., *New Essays*, trans. P. Remnant and J. Bennett (Cambridge: Cambridge University Press, 1981).

— *Philosophische Schriften* V (Frankfurt am Main: Insel Verlag, 1990).

Le Lionnais, F. (ed.), *Great Currents of Mathematical Thought*, trans. C. Pinter and H. Kline (New York: Dover, 1917).

Locke, J., *Essay*, ed. A. C. Fraser (Oxford: Clarendon Press, 1894).

MacIntosh, J. J. 'St. Thomas and the Traversal of the Infinite', *American Catholic Philosophical Quarterly*, 68 (1994), 157-177.

Maddy, P., *Realism in Mathematics* (Oxford: Clarendon Press, 1990).

Mancosu, P., *Philosophy of Mathematics and Mathematical Practice in the Seveenth Century* (New York: Oxford University Press, 1996).

Maor, E., *To Infinity and Beyond* (Boston: Birkhäuser, 1987).

McLarty, C., *Elementary Categories, Elementary Toposes* (Oxford: Clarendon Press, 1995).

Mill, J. S., *Autobiography*, ed. H. J. Laski (New York, 1952).

— *System of Logic* (London: Longmans, 1961).

Moore, A., *The Infinite* (London: Routledge, 1991).

Mordell, L. J., 'The Diophantine Equation $y^2 - k = x^3$', *Proceedings of the London Mathematical Society,* 13 (1913), 60-80.

Mordell, L. J., rev. of *Diophantine Geometry,* by S. Lang, *Bulletin of the AMS,* 70 (1964), 491-498

Ogilvy, C. S., *Excursions in Geometry* (New York: Oxford University Press, 1969).

Oresme, N., *Nicole Oresme and the Medieval Geometry of Qualities and Motions,* trans. M. Clagett (Madison: University of Wisconsin Press, 1968).

Pascal, B., *Oeuvres complètes* (Tours: Gallimard, 1954).

Penrose, R., *The Emperor's New Mind* (New York: Oxford University Press, 1989).

Perminov, V. Y., 'On the Reliability of Mathematical Proofs', *Revue Internationale de Philosophie,* 42 (1988), 500-508.

Plantinga, *Warrant and Proper Function* (New York: Oxford University Press, 1993).

- *Warrant: the Current Debate* (New York: Oxford University Press, 1993).

Plato, *The Collected Dialogues of Plato,* ed. Edith Hamilton and Huntington Cairns. Bollingen Series LXXI. (New York: Random House, 1961).

Popper, K. R., *Objective Knowledge* (Oxford: Clarendon Press, 1979).

Posy, C. J. (ed.), *Kant's Philosophy of Mathematics* (Dordrecht: Kluwer Academic Press, 1992).

Prawitz, D. (ed.), *Logic, Methodology and Philosophy of Science IX* (Amsterdam: Elsevier Science, 1994).

Prestet, J., *Nouveaux Elemens des Mathématiques,* 2nd edn. (Paris: André Pralard, 1689).

Reid, C., *Hilbert* (New York: Springer-Verlag, 1970).

Resnik, M., ed., *Mathematical Objects and Mathematical Knowledge* (Aldershot: Dartmouth, 1995).

Restivo, S., *Mathematics in Society and History* (Dordrecht: Kluwer, 1992).

Rota, G., rev. of *I Want to Be a Mathematician,* by Paul R. Halmos, *American Mathematical Monthly,* 94 (1987), 700-702.

Rotman, B., *Ad Infinitum* (Stanford: Stanford University Press, 1993).

Royce, J., *The World and the Individual* (New York: Macmillan, 1912).

Rucker, R., *Infinity and Mind* (Boston: Birkhäuser, 1982).

Russell, B., *Introduction to Mathematical Philosophy* (London: Routledge, 1919).

— *Mysticism and Logic* (London: Longmans, Green and Co., 1925).

— *Our Knowledge of the External World* (London: George Allen & Unwin, 1914).

Ryle, G., *The Concept of Mind* (London: Hutchinson, 1949).

Schattschneider, D., 'In Praise of Amateurs', in *The Mathematical Gardner,* ed. D. A. Klarner (Boston: Prindle, Weber & Schmidt, 1981).

Sénéchal, B., *Géométrie Classique et Mathématiques Modernes* (Paris: Hermann, 1979).

Sesiano, J., 'On an algorithm for the approximation of surds from a Provençal treatise', in *Mathematics from Manuscript to Print,* ed. C. Hay (Oxford: Clarendon Press, 1988).

Shapiro, H. N., *Introduction to the Theory of Numbers* (New York: John Wiley, 1983).

Smith, D. E., *History of Mathematics* (New York: Dover, 1958).

— *The Teaching of Elementary Mathematics* (New York: Macmillan, 1907).

Sorabji, R., 'Atoms and Time Atoms', in *Infinity and Continuity in Ancient and Medieval Thought,* ed. N. Kretzmann (Ithaca: Cornell University Press, 1982).

— *Time, Creation and the Continuum* (London: Duckworth, 1983).

Steen, L. A. (ed.), *Mathematics Today* (New York: Springer-Verlag, 1978).

Stillwell, J., *Mathematics and Its History* (New York: Spriner-Verlag, 1989).

Sweeney, L., *Divine Infinity in Greek and Medieval Thought* (New York: Peter Lang, 1992).

— *Infinity in the Presocratics* (The Hague: Marinus Nijhoff, 1972).

Thomson, J. F, 'Tasks and Super-Tasks', *Analysis,* 15 (1954), 1-10.

Toussaint, G., 'A New Look at Euclid's Second Propostion', *The Mathematical Intelligencer,* 15 (3) (1993), 12-23.

Tymoczko, T. (ed.), *New Directions in the Philosophy of Mathematics* (Boston: Birkhäuser, 1986).

Tzanakis, N., and B. M. M. de Weger, 'On the Practical Solution of the Thue Equation', *Journal of Number Theory,* 31 (1989), 99-132.

Vaihinger, H., *Philosophy of the As-If* (London: Routledge and Kegan Paul, 1935).

Voss, S., *What Number is God?* (Albany: State University of New York Press, 1995).

Whitehead, A. N., *The Aims of Education* (London: Williams & Norgate, 1950).

Wittgenstein, L., *Remarks on the Foundations of Mathematics* (Cambridge: The MIT Press, 1956).

Index

STUDIES IN THE HISTORY OF PHILOSOPHY